GLOBAL
WARMING
AND OTHER BOLLOCKS

GLOBAL
WARMING
AND OTHER BOLLOCKS

Professor **Stanley Feldman**
Professor **Vincent Marks**

JOHN BLAKE

Published by John Blake Publishing Ltd,
3 Bramber Court, 2 Bramber Road,
London W14 9PB, England

www.johnblakepublishing.co.uk

www.facebook.com/Johnblakepub facebook
twitter.com/johnblakepub twitter

First published in paperback in 2009
This edition published in 2014

ISBN: 978-1-78219-907-6

British Library Cataloguing-in-Publication Data:

A catalogue record for this book is available from the British Library.

Design by www.envydesign.co.uk

Printed in Great Britain by CPI Group (UK) Ltd

1 3 5 7 9 10 8 6 4 2

Papers used by John Blake Publishing are natural, recyclable products made
from wood grown in sustainable forests. The manufacturing processes
conform to the environmental regulations of the country of origin.

Every attempt has been made to contact the relevant copyright-holders, but
some were unobtainable. We would be grateful if the appropriate people
could contact us.

CONTENTS

SECTION 3: GRIDLOCKED BRITAIN:
A TRANSPORT POLICY?

SECTION 4: QUESTIONABLE DOGMA

AUTHOR PROFILES

EAMONN BUTLER is director of the Adam Smith Institute, an influential public-policy think-tank based in London. Non-party and non-profit, the Institute promotes free-market ideas and explores ways of introducing choice and competition into public services. Butler himself is a graduate of the University of St Andrews, and began his career working on policy issues for the US House of Representatives in Washington, DC. He is the author of books on famous economists including Adam Smith, Friedrich Hayek and Milton Friedman, and a book on how markets work, entitled *The Best Book on the Market*.

RODNEY CARTWRIGHT qualified in medicine from Birmingham University in 1963. After various clinical jobs, he specialised in medical microbiology that led to his appointment as director of Guildford Public Health Laboratory and a consultant microbiologist at the Royal Surrey County Hospital. He worked closely with the University of Surrey, where he became an honorary visiting professor. Cartwright developed special interests in the epidemiology and prevention of travel-associated infections and water-borne diseases and became a medical adviser to the tourist industry and many British water companies. He is still an independent medical adviser to the UK government's Regulator for Drinking Water Standards. His wider interest in public health resulted in his serving as a trustee and council member of the Royal Institute of Public Health and as Master of the Worshipful Company of Plumbers.

STANLEY FELDMAN gave up working for his biochemistry thesis on the metabolism of the wood louse to study medicine in 1950. Qualified (hons) Westminster Medical School 1955. Trained as anaesthetist at Westminster Hospital. Learned research techniques as a Fellow University Washington USA 1957–58. Senior Lecturer Postgraduate Medical School, 1962. Advisor in Post-Graduate Studies Faculty Anaesthetist 1965–70. Visiting Professor Stanford University USA 1967–68. Higginbotham Lecturer Dallas, Frederickson Orator Emory University. Member Senate University London, Chair of Anaesthesia University London at Charing Cross and Westminster Medical Schools (later Imperial College School Medicine). Research adviser Royal National Orthopaedic Hospital 1994–97. Author/Editor: 12 text books in anaesthesia, including *Scientific Foundations of Anaesthesia*, *Mechanism of Action of Drugs*, *Drugs in Anaesthesia*, etc. Editor *Journal, Anaesthetic Pharmacology*. Contributor *Encyclopaedia Britannica*. Published over 100 peer-reviewed papers on molecular mechanisms of drug action, the evolution of animal life and science education. Recent publication *From Poison Arrows to Prozac*, 2008, John Blake. *Panic Nation* (with Vincent Marks), John Blake, 2005. *Life Begins*, 2007, Metro Books.

DAVID JEFFERS began his engineering career as a college apprentice with Metropolitan Vickers and he was working on nuclear power station construction in the heady days of the early 1960s, when nuclear electricity was going to be too cheap to meter. After Berkeley Power Station was commissioned, he joined Manchester University as reactor engineer on the joint Manchester and Liverpool Universities' research reactor. Jeffers completed a PhD in nuclear engineering while working there but, after this period in academe, industry beckoned again and he transferred to the Central Electricity Generating Board's (CEGB's) Scientific Services Department, which provided technical backup to the power stations in the Northwest. He became controller of scientific services in the Northwest and then director of design in the CEGB's Transmission Division. Between the privatisation of the industry and his retirement in 1999, he was the National Grid Company's consultant on electric and magnetic fields.

VINCENT MARKS graduated in medicine from Oxford in 1954. He took up a consultant post in Epsom at one of the largest clinical laboratories in the Southeast of England, where he established the first drug-screening laboratory in the country in addition to continuing his clinical work into the study of metabolic disorders. He moved in 1970 to the University of Surrey in Guildford as its first professor of clinical biochemistry and subsequently became its first dean of medicine. Marks is known internationally for his research on insulin and hypoglycaemia. He has published numerous books and monographs including *Insulin Murders* and *Panic Nation* for non-specialists.

MICHAEL SMITH, emeritus consultant physician to Royal Surrey County Hospital, Guildford, qualified in medicine in Newcastle in 1957 and held junior posts in professorial departments of paediatrics and medicine and a year with Dr Henry Miller in neurology. After two years of national service with the RAMC, he returned to Newcastle as registrar and then was appointed as research registrar at the Royal Postgraduate School of Medicine. Smith spent a year as research fellow at the University of California, Los Angeles, developing an interest in diabetes, endocrinology and metabolic medicine. He was appointed consultant physician in Guildford. Undergraduate and later postgraduate tutor and regional adviser in postgraduate education. Honorary senior lecturer in biochemistry, University of Surrey.

ANDREW TAYLOR is a consultant clinical scientist in the Clinical Biochemistry Department at the Royal Surrey County Hospital in Guildford and an honorary reader at the Faculty of Health and Medical Sciences of the University of Surrey. He is the director of the Supra-Regional Assay Service Trace Elements Centre at Guildford and has more than 30 years' experience with the clinical biochemistry and measurement of trace elements in biological and environmental samples. Taylor's research interests have involved the clinical biochemistry of mercury, gold, aluminium and lead, and current work includes the release of metals from prosthetic implants. Improvements in measurement techniques have been another interest and he established a proficiency testing scheme to monitor performance of analytical laboratories. This now involves collaboration with the organisers of similar schemes in other countries to ensure international consistency of results.

AUSTIN WILLIAMS is an architect and project manager. He is the Director of the Future Cities Project; past editor of the *Architect's Journal*. He is the author of *Towards a New Humanism in Architecture* and more recently of *Enemies of Progress*, a book on the effects of sustainability. He is the producer and chair of the Bookshop Barnies and a contributor to the Battle of Ideas.

PAUL WITHRINGTON is the director of Transport Watch. He has an honours degree in civil engineering from Bristol University and a masters degree in transport planning, having graduated in 1967. He is a member of the Institution of Civil Engineers (a chartered engineer). As a transport planner he first worked for the Greater London Council (1967 to 1970). At that time four ring roads of motorway proportions were planned for the capital. It was in that organisation that he had contact with the Railway Conversion League. Subsequently, he lectured at Portsmouth Polytechnic, had experience with a small PTRC, a firm that specialised in providing a range of courses for local and central government and in 1975 he joined Northamptonshire County Council, where he worked for 20 years as the project manager of Transport Planning Department. He retired in 1994 to work as the director of Transport Watch.

THE DOOMSDAY VIEW
OF THE WORLD

THE IDEA THAT MAN is born into sin is as old as the story of Adam and Eve. It spawns the doomsday cults and the belief that only through repentance and self-denial can man aspire to redemption and save the world. It is manifest in the abstinence of Lent, the fasting of Yom Kippur and Ramadan and the various New Age groups that deny themselves material pleasures and seek solace and forgiveness by adopting a primitive lifestyle that they claim is compatible with nature. This has given rise to the 'back-to-nature cults' that predict that man's tampering with nature will lead to the end of the world as we know it. In the early days it spawned the Luddite movement and opposition to mechanisation of transport and the telephone. Today it sees industrialisation, modern farming methods, genetically modified crops, scientific medicine, nuclear power and the internal-combustion engine as heralding the end of civilisation.

Fortunately, all the end-of-the-world cults have been proved wrong, but, undeterred by this abysmal record, they continue to echo a deep-seated, inherent fear that, unless man repents and changes his way of life, he will be instrumental in destroying his world.

The aesthete, the hermit and the back-to-nature movement all reflect this mindset. Today, politicians and persuasive pressure groups play on this same basic fear. They scare us with tales of an inevitable catastrophic effect of global warming, which they blame on the carbon dioxide (CO_2) we produce; they play on a fear that an epidemic of obesity, caused by overeating, will kill our children before they reach a mature age; they claim that our indulgent lifestyle is consuming Earth's precious resources, that pesticides will kill off life in our oceans, that chemicals in food will poison us all and that invisible rays from nuclear power stations, overhead electric cables or mobile phones will kill us with cancer. They are today's prophets of doom using modern media methods to disseminate fear so as to propagate their belief that the end of the world is nigh. This book examines the evidence for and against some of their apocalyptical predictions.

FOREWORD

The philosophies of one age have become the absurdities
of the next, and the foolishness of yesterday becomes
the wisdom of tomorrow.

—Sir William Osler,
Aequanimitas and other addresses

SEVERAL YEARS AGO we were told that all the computer systems would fail at the turn of the millennium. Millions of pounds were spent on the so-called 'millennium bug'. It proved to be nonsense.

The panic caused by public concern over the toxicity of DDT and the prediction it would lead to a sterile sea and famine on Earth caused its use to be suspended. As a result, millions have died of malaria and insect-borne diseases.

The health authorities told us that we should avoid cholesterol, or else our arteries would clog up and we would die young from heart disease. Eat only one egg a week, drink only skimmed milk and consume no cream or butter. Millions of pounds were spent on low-cholesterol diets. We now know it is a waste of time: it is not the cholesterol one eats that causes heart disease.

We are told our children are getting so fat that they will die younger than their parents. It is nonsense. We are being scared stiff.

Following two relatively dry years in 2004–05, we were told by government officials that, because of global warming, we should get used to droughts and obtaining our water from standpipes in the streets, that we should rip out our flowering shrubs and replant with drought-loving plants. So successful was this propaganda that the Chelsea Flower Show in 2007 featured grasses and succulents. It was followed by a series of the wettest summers on record, with sodden gardens and floods.

Today our children's school meals are subject to the 'food police', who inspect their sandwiches, control their lunches and lecture them on their parents' bad eating habits and the dangers of junk food. They ban fizzy drinks and stop the children using salt. They are taken on 'school trips' to the depots to be indoctrinated on the virtues of recycling, shown propaganda films to scare them stiff about the dangers of global warming and told they must walk to school and stop their parents using the car and destroying the Earth. They are not allowed to expose themselves when the sun shines in case they get melanoma, they must avoid foods that contain chemicals and weigh themselves regularly to make sure they are not part of the obesity epidemic that will kill them before they reach maturity.

With such a busy schedule, lessons on reading and arithmetic take second place in the curriculum. This might possibly be justified if the message they are being given was likely to improve their wellbeing and prevent future disease. The trouble is that there is increasing evidence that it is, like the 'millennium bug', a belief that is in part or entirely untrue. Often it is based on numbers picked out of the air.

We are frequently presented with new scare stories, based on so-called 'new research', often involving epidemiological surveys, poor science or inadequate samples. They soon become belief systems. They are propagated by the prophets of doom in a manner that engenders irrational fear. They foretell of imminent disaster unless remedial action is taken. The scare develops into a belief, like a religious dogma, whose tenets invariably end up involving a restriction of personal choice and freedom of action. This, they claim, is the price to pay to assuage the wrath of nature and the inevitable disaster presaged by these New Age prophets of doom.

The problem is that not all scare stories are baseless or involve poor science. Smoking is undoubtedly bad for your health. The beneficial effect of stopping smoking is real and important. The smoking scare was based on good evidence that met the criteria of 'reasonable proof'. Similarly, we have reason to be circumspect about the effect of a burgeoning population on finite or limited resources and to be prepared to meet the challenge of any possible natural disaster or epidemic. It needs good evidence from reproducible scientific observation to determine the degree of risk associated with a particular indulgence. Only when this evidence is watertight and compelling is it justifiable to restrict personal freedom and individual choice.

The various chapters in this book analyse some of the beliefs and opinions that are the basis of the constraints that have been imposed on the way we live. In some cases, our analysis justifies the measures that have been taken. In most cases, they challenge the accepted dogmas that have led to the limitations that have been imposed on our freedom. The particular views presented reflect the opinions of the authors. Their views are personal and are iconoclastic. They are not necessarily shared to the same extent by all the contributors or the editors. They are intended to make the reader stop, enquire and think again about what he or she believes to be an accepted fact.

INTRODUCTION

STANLEY FELDMAN

*Knowledge may be expensive but it is
much cheaper than ignorance.*

*The pragmatist says, 'I change my mind when the facts change.'
The dogmatist says, 'My mind is made up – do not
confuse me with facts'.*

NEWSPAPERS, magazines, television and radio are constantly
bombarding us with stories presented as facts. Unfortunately, many
of them turn out to be opinion rather than something that has been
proven to be likely or correct. The problem we face is how to
distinguish between a belief or an opinion, and a fact.

In the courts of law it is the accumulation of evidence that helps
to prove a proposition 'beyond reasonable doubt' in criminal cases,
or, in the civil courts, the lower standard of proof, 'on the balance
of probabilities'. In the world of science the same criteria should
hold true. It requires good, reliable evidence to substantiate a
hypothesis. All too often the stories that reach the media fall far
short of the level of proof required by the criminal courts and
seldom even reach that required by the civil courts. Without
corroboration all information is uncertain and exists in a twilight
world of half-truths and make-believe.

In his book *The Devil's Chaplain*, Richard Dawkins, the professor of public understanding of science at Oxford University, wrote a letter to his daughter in which he says that it is his greatest wish that when she is grown up and is told that something is a fact she should ask herself, 'Is this a thing people know because of evidence?'

Dawkins wanted to impress upon his daughter the importance of distinguishing between dogma – or firmly held belief – and verifiable fact. Unfortunately, it is often more comfortable not to try to make the distinction but to 'go with the herd' and unquestionably accept what one is told because it is believed by so many seemingly rational and knowing people. As the author Michael Crichton pointed out, science is based on evidence and proof – not popular opinion or 'consensus'.

Let's be clear: the work of science has nothing whatever to do with consensus. Consensus is the business of politics. Science, on the contrary, requires only one investigator who happens to be right, which means that he or she has results that are verifiable by reference to the real world. In science consensus is irrelevant. What is relevant is reproducible results. The greatest scientists in history are great precisely because they broke with the consensus. There is no such thing as consensus science. If it's consensus, it isn't science. If it's science, it isn't consensus. Period.

Accepting a belief as a fact, without convincing evidence, is the way to become a member of an unthinking society of anti-intellectual zombies. Throughout the history of civilisation, many ideas that were previously held as certainties, supported by belief and experience, have been found wanting in the light of new information. Newtonian mathematics, which formed the basis of so much of our teaching, has been found to be only half true by the modern theories of quantum mechanics.

Our knowledge of the universe has been demonstrated to be based on inadequate concepts that were taken as fact before the theory of relativity and the present means of exploring space had shown their limitations. Until recently we were told, with absolute certainty, that foods rich in cholesterol caused heart disease. This is now known to be wrong. We are constantly developing new ways of testing and evaluating evidence that is causing us to question previously accepted dogma. This is the basis of knowledge; it is the way of science.

To meet these uncertainties we need to scrutinise beliefs and examine past evidence in the light of new knowledge. It is by experimental evidence and by reason that we distinguish between dogma and fact. Just as the availability of new methodologies, such as DNA typing in criminal investigations, has led to examination of past convictions, and space exploration has shed new light on the origin of the planets, so new research will invariably cause us to challenge more and more present 'certainties'. If further research casts doubt on the original propositions, it must be categorised as a hypothesis, not as fact. If the new evidence – provided it can be substantiated – is incompatible the hypothesis must be abandoned. To persist in maintaining doubtful dogma in the face of evidence, puts it in the realm of an unsubstantiated 'belief' or myth.

In this book we have looked at some commonly held beliefs – dogmas that countenance no challenge either because they are taken for granted by those in authority or they have the seal of, so-called, 'scientific approval'. While it is often more comfortable to accept some of these dogmas – indeed to challenge them one runs the risk of being labelled an iconoclast – it is our duty to encourage

INTRODUCTION

debate, however certain we are of our own correctness. It is necessary to examine all of the evidence constructively to find out how much supports and how much contradicts a particular proposition or hypothesis. For there to be confidence in a hypothesis it is essential that any evidence that questions the validity of its assumption be answered. When they are, it can be dignified by its description as a theory.

Belief, hypothesis and fact

Many people believe in a deity of some form. However, God is not subject to evidence or proof. That so many believe in his/her/its existence does not make God a fact. The absence of evidence of there being a god, or the impossibility of ever obtaining proof of his/her/its existence, distinguishes theism as a belief from observable, confirmable fact. Belief in God is neither a theory nor a fact: it is an unsubstantiated hypothesis based on faith. It may bring immense comfort to those with it but it does not necessarily make it true.

The purpose of this book is to make the reader think again about what he/she is told is a fact. Too much of the information we are given turns out to be beliefs or hypotheses dressed up as fact. When 'everybody knows' something it's time to challenge it; the 'scientists believe' phenomenon, especially when supported by statements such as 'new research has shown', should always arouse suspicion. The use of emotive, often pejorative, terms such as 'will cause the end of the world', 'will kill hundreds of thousands' and 'time is running out' should alert readers to the probability that they are being mentally bullied and manipulated. It should warn them of the need to examine the evidence carefully to see whether the case is indeed proven 'beyond reasonable doubt'.

Defining our terms

Beliefs are authoritative statements that have not been subjected to testing or are incapable of being tested. They rely on anecdotal evidence and the power of authority.

Hypotheses are tentative beliefs that are subject to proof or disproof and consequently a step beyond 'belief'.

Facts are measurable and reproducible events. There is no known or current evidence that casts doubt upon their truthfulness. They are therefore 'beyond reasonable doubt', although new evidence may emerge to question their accuracy in the future. Such is the nature of progress.

To illustrate the importance of separating these three categories let us consider some examples.

Before the 16th century it was universally believed that the sun orbited the Earth. This was a consensus opinion. Not to accept this idea was regarded as heresy. It was an authoritative dogma promulgated by the Church and could not be questioned. The authority of the Bible was invoked as evidence to support the claim – if, indeed, evidence was required, since it was 'self-evident' from everyone's personal daily experience. The idea that the Earth spun at an incredible speed as it orbited the sun seemed as improbable as it was blasphemous. Unlike many beliefs, this one was subject to scientific evaluation. Nevertheless, it was some time after Copernicus and Galileo had produced convincing evidence that the Earth orbited the sun, rather than the more facile explanation, that the heliocentric view of the world became universally accepted.

Today, many people buy 'organic food' in spite of the extra expense because they believe it is 'better' for them than food grown or produced 'conventionally'. There is no evidence to support this belief. Its supporters have been persuaded by unsubstantiated propaganda, repeated by what they see as authoritative sources, such as Prince Charles, the Soil Association, many retailers, celebrity chefs and some self-serving nutritionists. They ignore the evidence and remain unconvinced by conclusions reached by bodies they would normally be expected to respect, such as Food Standards Agency, the Advertising Standards Authority and highly regarded nutritionists who have contributed to and examined the evidence. This has consistently failed to show any advantage in 'organic produce', which is in effect a trademark rather than a genuine entity. Insistence on buying organic food is based on belief, not on evidence, and is as irrational as the belief that the world is flat.

The germ theory of disease

Before the pioneers of microbiology proved otherwise, many infections were believed to be due to draughts, exposure to damp or cold – perpetuated in the term 'catching a cold' – or a miasma emanating from swamps (malaria). It was the perfection of better optical lenses for microscopes that led to the identification of bacteria and other microscopic organisms as the cause of these infectious diseases. Identification of viruses came later with invention of the electron microscope.

The 17th–18th-century German physician Robert Koch, who identified the tubercle bacillus and proposed that it was the cause of tuberculosis, was disbelieved by many of his contemporaries. In order to prove his hypothesis Koch established three most important elements of proof of causation of disease by a toxic agent.

Now known as *Koch's postulates*, these are:

1. the proposed cause of the disease must always be shown to be present in a person suffering the disease;
2. removing the proposed cause cures the disease; and
3. reintroducing the putative cause re-establishes the condition.

Koch went on to fulfil these three requirements for proof in experimental animals. As a result of this evidence his hypothesis became a fact. 'The Germ Theory of Disease', to which Koch's and others microbiologists' work had given rise, was espoused by many practitioners as though it applied to all diseases whether caused by micro-organisms or not. Clearly, it is going to be difficult, and often ethically improper, to try to satisfy all three of Koch's postulates in all cases, but the more that are met, the more likely the proposed causal mechanism is to be correct. A recent example to hit the headlines, but one that fulfilled none of Koch's postulates, was the suggestion that the measles virus is a cause of autism.

Causation versus coincidence

Illnesses caused by exposure to noxious substances, such as arsenic, have long been recognised. Those that take many years of exposure before the illness becomes apparent are much more difficult to identify.

Richard Doll, the 20th-century British epidemiologist, demonstrated a relationship between cigarette smoking and the risk of developing cancer of the lung. He suggested that smoking caused cancer of the lung after he had surveyed people's smoking habits and found very little effect in those smoking fewer than five cigarettes a day, whereas above this level the frequency with which they developed cancer increased with the number of cigarettes they smoked. This dose-related phenomenon added considerable credibility to his findings. When he further demonstrated that people – doctors, it so happened – who voluntarily stopped smoking reduced their risk of cancer dramatically, he had sufficient evidence to convince all but the most sceptical of a causal rather than chance association between the two. Without the supporting evidence the relationship between smoking and cancer of the lung would have remained a tentative hypothesis.

Failure to seek evidential confirmation can produce half-baked theories, often presented in the media as 'new scientific evidence', that can be very misleading.

Some years ago there was an epidemiological study that showed that the residents of the island of Okinawa, in the Pacific, lived longer on average than those on neighbouring islands. It was claimed that this was due to their high consumption of yams. The researchers clearly demonstrated that they did eat more yams and they did live longer. To have turned this hypothesis into proof they would have needed to show that the lives of the residents of Okinawa were shortened if they *gave up* eating yams and that by eating yams those living on nearby islands increased *their* life expectancy. They did not do this and the hypothesis remains unproven. It is highly probable that the conclusions are wrong and that the observations are better explained by the high proportion of the inhabitants of Okinawa who are of Japanese descent. Japanese have, in general, a longer expectation of life than almost all the other people living in the Pacific.

Just as important as producing evidence to support a particular theory is the ability to demonstrate that any alternative explanation is wrong. It was this concept that led the distinguished 20th-century philosopher Karl Popper to propose that science advances by *dis*proof. He pointed out that, if you made the hypothesis that 'all swans are white', it did not matter how many white swans you found, or how much evidence you gathered to support your theory, as soon as a *black* swan was found, the theory becomes untenable. However, it is always possible that the black swan was a white one that had been painted black or an odd genetic freak.

In order to demonstrate the validity of the disproof, it would be necessary to show that it *was a swan* and that it *was naturally black* and *bred other black swans*. It is just as important to support any disproof with evidence as it is for proof. Popper pointed out that disproof is much more potent as an evidential weapon than proof. One element of disproof will dispel ideas supported by volumes of positive evidence. A tablecloth with a dirty stain in one corner is 'dirty' regardless of the fact that 99.9% of it is clean!

This book

This book is about evidence. It challenges some of the assumptions that we have come to accept as fact. It presents evidence that either supports the hypothesis or indicates that it is not necessarily the only, or even the best, explanation of observed events. In some instances, even the best analysis of present dogmas is unable to answer concerns about future uncertainties. Austin Williams points out in this book how mankind has, in the past, survived by overcoming the problems posed by nature while the doomsday soothsayers advocate a retreat into a primitive world of doing less, having less and achieving less. On the other hand, few would question the view put forward by Vincent Marks, in his chapter on population, that there must ultimately be a limit to unbridled population growth and consumption. However, no one knows what that limit is or how it will be affected by new technology and by world events.

We live in an uncertain and unpredictable world – but none of us enjoys uncertainty. Nevertheless, we have to make decisions, and do so on the 'balance of probabilities' when we know what the alternatives are. Some scares, such as the probability that we will suffer a human bird flu pandemic sometime in the future, are justified by scientific evidence; some, such as the effect of unlimited growth, depend upon unpredictable future developments; while yet others are patently wrong. The millennium-bug scare was groundless. So were prophecies that AIDS and new-variant Creutzfeldt–Jakob disease (CJD) would decimate the population. Similarly, the prediction, made in the 1970s, of the coming of a new ice age and the authoritative forecast in 2006 that rainfall in England would decrease to a point where drought would be usual. All were pronouncements that made headline news but all have proved to be wrong.

The paradox that many of the world's great scientists are or were (religious) believers will not have escaped our readers' attention. The ability to accommodate a belief system at the same time as a rigorous scientific inquisitiveness, in the same brain, is far from rare. Few things in life are certain beyond having a beginning and an end – and some even question this.

In our present society, government bodies and authorities increasingly make decisions for us about the kind of life we live. It is essential, therefore, that their recommendations are based on sound evidence. Too often they are the result of an ill-conceived overreaction to pressure by special-interest groups, a media campaign or zealots presenting their beliefs in the guise of fact. As a result, the way we live, what we eat, what we do and how we spend our money is often based on a doubtful dogma that originates from often well-meaning government circles. This can be dangerous, as it undermines the influence of credible government warnings and advice, such as the benefits of using seat belts in cars, and dangers of drinking excessively or of smoking.

Good advice is helpful and should be welcomed, but it must be based on evidence and not on questionable dogma. It should be permissible for governments to admit that the evidence upon which they have to base policy is at the civil rather than criminal level of certainty and often not even as good as that.

1

CO$_2$ AND GLOBAL WARMING

GLOBAL WARMING

STANLEY FELDMAN

We have to ride the theory of global warming, even if it is wrong.
— Timothy Wirth, ex-president,
United Nations Foundation

Solutions to climate change must be based on firm evidence, not dubious ideology… Policies must be based on science rather than the dogma of the environmentalist movement.
— Pope Benedict XVI,
December 2007

DOGMA
Manmade CO_2 is causing global warming,
which will cause catastrophe.

SINCE THE INVENTION of the telescope, the possibility that there was life on other planets has excited astronomers. As we learned more about our galaxy we soon came to realise that most planets are such inhospitable places that life, in any recognisable form, is improbable. One of the most compelling arguments against there being life on most of the planets is the extremes of temperatures that occur on their surface. When they are exposed to the full effect of the sun the temperatures soar, but once the sun sets the cold is so intense that most forms of life would freeze.

The reason why Earth is not subjected to these extremes of temperature is the presence of its peculiar 'atmosphere', which provides a protective blanket of gases, containing nitrogen, oxygen, water, argon and a tiny amount, 0.038 per cent, of carbon dioxide. Without this atmosphere the average temperature on the planet would be about minus 18°C. It is thanks to their effect that our planet is habitable.

It was the observations of the French mathematician Baron Joseph Fourier in 1822 that led to our understanding of the importance of the atmosphere in making our planet habitable. He suggested that it was the presence of particular gases in the atmosphere that moderated the extreme heat produced by the sun. Some years after Fourier, John Tyndall – the Irish physicist who described the Tyndall effect caused by refraction of light – demonstrated that, of the various gases in the atmosphere, only a few of them had a significant effect in preventing the full force of the sun's energy reaching the surface of the Earth, causing it to become unbearably hot and, by preventing the warm surface losing its heat at night, they prevented it from freezing. He suggested that these gases captured and stored the sun's energy that made the Earth warm. As a result, they acted as a buffer for solar energy. He studied some of the gases that occur in the air and found that, of the gases he investigated, CO_2 was the most important buffer.

In fact, molecule for molecule, water is a much more potent greenhouse gas than CO_2, since it absorbs energy over a far wider energy-wave spectrum. One has only to consider the effect of supplying energy, in the form of heat, to water in a kettle, or in the form of microwaves in an oven, to appreciate its ability to absorb energy. It is this energy-absorbing property that makes it useful in dowsing fires.

Methane is also about 20 times more potent than CO_2 as a buffer of energy because it absorbs energy over a larger energy-wave spectrum than CO_2, but it is present in only minute amounts in the atmosphere and its overall contribution to this effect is very small. In spite of their high concentrations, nitrogen and oxygen do not absorb energy in the infrared energy spectrum and do not contribute significantly to this buffering effect.

Birth of the greenhouse analogy

It was the Swedish polymath Svante Arrhenius (1859–1927), who in his dissertation in 1896, 'The influence of Carbonic Acid in the Air on the Temperature on the Ground', first used the greenhouse analogy. He prepared the way for our understanding of the 'greenhouse-gas effect'. In his experiments he confirmed the importance of CO_2 in preventing the Earth from cooling rapidly when the sun goes down. He believed the effect was so powerful that, as a result of the large amount of wood burned in winter in Sweden, sufficient CO_2 would be given off to increase the greenhouse effect and to produce a warm, productive climate all the year round. Using the Stefan–Boltzmann equation, he calculated that doubling the amount of CO_2 in the atmosphere would increase the temperature on the ground by an overall 4–4.8°C. At the time he made his observations there was no way of measuring the concentration of CO_2 at the very low levels present in the atmosphere. As the actual concentration of CO_2 is less than 0.038

5

per cent, it is easily doubled by what, in global terms, is a very small increase in CO_2. It is the fear that in the next hundred years human activity will add significantly to the CO_2 concentration in the atmosphere that has provoked the present global-warming panic.

Although the term 'greenhouse-gas effect' has come to be used to describe the total effect of all elements in the atmosphere on the temperatures we enjoy on the surface of our planet, it was only the part played by CO_2, methane and sulphur dioxide, in preventing Earth from freezing when the sun sets, that was highlighted by Arrhenius.

The sun is the only major source of heat on our planet. When it shines it warms us up. Some of that energy is absorbed by the Earth's crust, warming it up. Some of that heat diffuses from the surface into the soil and rock below. This is *geothermal heat* and it can be tapped into by the heat pumps that are used to supplement domestic heating. As the surface of the Earth gets hotter, it acts like a radiator, dispersing its heat back into the atmosphere. Because this radiation is in the infrared, invisible spectrum, we cannot see it happening, although we can feel its effect as it warms us up. It is the infrared radiation given off by the planet once it has been warmed that is buffered by the greenhouse gases in the atmosphere. They do it by absorbing the energy within their molecules. It was Arrhenius who pointed out that it was those gases that had the capacity to absorb energy with a wavelength in the infrared spectrum that could buffer heat energy in this way. He studied CO_2, methane and sulphur dioxide, all of which share this property. However, it is water molecules that are the most potent of the greenhouse gases because they absorb energy over a very wide range of wave bands, from infrared to visible light.

Were it not for the blanket of greenhouse gases in the atmosphere the temperatures on the earth would be boiling hot during the day and freezing at night. It is the result of the presence of these gases in our atmosphere that some of the infrared energy that would otherwise reach the surface of the Earth when the sun is shining is absorbed, reducing the extreme heat that would otherwise ensue.

Because of the greenhouse gases, much of the Earth's warmth, acquired during the day, is prevented from being dissipated into space at night. The proponents of the theory of anthropogenic global warming (i.e. global warming caused by humans) point out that, as a result of this effect, increasing the amount of greenhouse gases in the atmosphere will reduce the amount of heat dissipated into space at night and therefore, over time, it is likely to cause the earth's temperature to rise.

Arrhenius failed to persuade the climatologists of his era of the significance of his theory. The wood burning in Sweden did not produce the warm winters he forecast. His predictions, based on the greenhouse-gas effect of CO_2, turned out to be wrong. The theory did not explain the recurrent swings in temperature that have occurred since the Earth was born. It failed to answer the question of why, in the not-too-distant past, his Scandinavian ancestors had been able to live, farm and colonise Greenland and the frozen north of his country but had been forced to abandon their settlements due to the encroaching ice in the 16th and 17th centuries, without evidence to suggest there had been any reduction in the atmosphere's CO_2 concentration. It seemed more likely, to the climatologists, that the explanation lay with the small deviation that occurs in the axis of Earth as it circles the sun and the flattening of its slightly elliptical circuit, the so-called Milankovitch effects.

A new ice age?

In 1938, a British scientist, G S (Guy) Callendar, drew attention to the potential beneficial effect on the agriculture and farming of a rise in temperature as a result of an increase in CO_2 in the atmosphere, but his views were largely ignored. His predictions were based on the observations of Dr Keeling's team of scientists on the CO_2 content of the atmosphere at the top of a mountain in Hawaii. They had demonstrated that the level of CO_2 in the atmosphere in the Pacific was slowly rising each year. Callendar's words went unheeded as, in spite of his prediction of a warmer world, it was followed by a 35-year period, from 1940 to 1975, of progressively colder times. Before we accept that an increase in CO_2 causes warming of the planet, the fall in temperature that occurred in this period in spite of a documented incremental rise in CO_2 must be explained.

It is almost certain that the levels of atmospheric CO_2 would have increased rapidly during World War II and the post-war reconstruction period, and that this should have caused a rise in temperature due to the greenhouse-gas effect. But the temperatures actually fell by 0.3–0.4°C. The effect was sufficiently worrying for climatologists, at that time, to warn of the advent of a new ice age.

2

GREENHOUSE GASES

STANLEY FELDMAN

DOGMA
Greenhouse gases are all the fault of human activity.

ANYONE LYING ON an English beach enjoying the summer sunshine could be excused for jumping to the conclusion that it was the occasional cloud that obscures the sun that causes their world to cool. They are probably right. Under cloudless skies the temperature drops dramatically the moment the sun sets or one moves from the sunshine into the shade, although the CO_2 levels do not change.

When one listens to the weather forecast, it is clear that it is the cloud cover that determines whether the sun will shine and the weather will warm up. It is the amount of energy in the water molecules that make up the clouds that determines whether or not a particular wind will warm us up or cool us down. It would be nonsense to consider the effect of gases in the atmosphere on Earth's temperature without accepting a major role for these clouds, especially those at the lower levels. To concentrate solely on the CO_2 ignores the fact that many different processes are involved in determining the planet's temperature.

There is one principal source of heat, and that is the sun; it far exceeds any other influence on the global temperature. Nevertheless, significant but comparatively small amounts of geothermal energy are constantly being released from the molten mass in the depths of the Earth. This energy warms areas of the oceans, through the hydrothermal vents of underwater volcanoes, and the land where the Earth's crust is sufficiently thin to allow thermal warming, as in, for example, Arizona, Iceland, New Zealand and Antarctica. However, their global contribution of energy is tiny.

There is good evidence that the sun's heat varies from time to time and that this is related to magnetic activity and sunspots. There is strong correlation between the number and frequency of these changes and the sun's energy output. When there are a lot of sunspots the energy is reduced and the temperature, not only of Earth but of other planets such as Jupiter and Mars, falls slightly. Sunspots affect not only the amount of heat given off by the sun but also the amount of cosmic bombardment from outer space, due to their strong magnetic effect.

The mean temperature on Earth depends upon how much of the sun's energy reaches the surface of our planet when the sun shines and how much of this heat is lost from the Earth when it gets dark. There is no doubt that this is affected by the atmosphere.

Over the past century, solar irradiance has increased, which in itself would account for a 0.2°C rise in surface temperature if no other mechanism existed to affect the transfer of this energy from the sun to the Earth. However, various factors in the atmosphere affect this process.

The most obvious is the effect of the clouds.

The importance of water

Clouds are composed of water vapour and droplets; the higher the concentration of water droplets, the darker and more thunderous are the clouds. By and large the lower the clouds, the higher the concentration of water droplets. When a cloud appears to obscure the sun it does so by reflecting the sun's energy, including that in the visible spectrum, back into space, so that its light fails to reach us on Earth. This reflective action depends largely on the concentration of water as droplets. Because molecules of water also absorb the warming, shorter-wave, infrared energy, we also lose much of the sun's heat when it is cloudy.

The water molecules in the clouds also affect the surface temperature of the Earth in their role as a greenhouse gas. They blanket over the Earth and prevent the escape of infrared energy from its surface, stopping it cooling. That is why cloudy nights are much warmer and balmier than clear nights. On cloudless nights the temperatures tend to drop rapidly once the sun sets as the atmosphere lacks the greenhouse effect of the water in the clouds. It is evident on these occasions that our comfort depends to a much greater extent on the water in the atmosphere than it does on CO_2.

We are coming to realise that clouds themselves, especially their disposition and composition, are also affected by solar activity, although the contribution this makes to the temperature of our planet is difficult to quantify.

11

The concentration of water in the atmosphere varies. Even on a 'dry day' the air we breathe is moist and the air we exhale is saturated. If one looks at the amount of ice deposited in the freezer compartment of a refrigerator it is evident that the air in the refrigerator, which may have appeared to have been dry, in fact contained a lot of water, some of which was deposited as ice when it was trapped inside the refrigerator when the temperature fell.

Greenhouse gases, like carbon dioxide and water, merely store part of the energy that originated in the sun. They absorb some of the energy that radiates from the sun when it shines and from the Earth after it has been warmed by the sun. At night CO_2 and water vapour act together as a blanket over the Earth, minimising the loss of heat as the atmospheric temperature begins to fall.

The effects of the various components involved in determining the Earth's temperature are difficult to separate quantitatively. The overall effect of the clouds is especially difficult to measure, as it varies enormously from time to time and from place to place. Its effect on the sun's energy depends upon whether it is present as a vapour or as droplets. If all the water in the atmosphere acted effectively as a greenhouse gas it would contribute a massive 96 per cent to this effect, dwarfing any contribution from CO_2.

It is generally agreed that the water vapour and droplets in the clouds contribute, at the very least, 50 per cent of the total global warming effect, although many others suggest that the figure should be much nearer 93 per cent. The trouble is that the amount of water vapour in the atmosphere varies from time to time, and it is impossible to predict.

A year without summer

A lot of polluting particles in the atmosphere will enhance the effect of the clouds in insulating Earth from the effect of the sun. It is believed to be responsible for the absence of significant global warming in southern China and parts of India over the past 15 years (2008 saw their coldest winter for 50 years). Volcanic eruptions spew out polluting particles (as well as CO_2 and other greenhouse gases), which influence the climate. The Tambora volcanic eruption of 1815 produced a year without a summer due to the water vapour and the particles released when it erupted. They formed a cloud over large parts of the planet. The Krakatoa volcanic explosion of 1883, which released huge amounts of debris, CO_2 and sulphur dioxide into the atmosphere, affected the temperature of the world for two to three years. Although CO_2 and sulphur dioxide are greenhouse gases, the world cooled.

As a result of the reduction in the concentration of particle pollutants that used to hang in the air over London, before the Clean Air Acts of the 1950s, the sun's rays are now better able to penetrate the atmosphere. This has led to an average increase in temperature in the city of about 2°C. This is as great as the increase in temperature predicted in the next hundred years by some of the models of global warming! Together with urban warming, this has resulted in the temperature in central London being up to 4°C warmer than the neighbouring countryside.

Scientific controversy centres on the relative importance of the enormous, but almost impossible-to-measure, effect of water vapour and clouds, the effect of changes in the sun's activity and the effect of the tiny but increasing 0.038 per cent of CO_2 in the atmosphere, in the production of global warming. It is because these effects are impossible to separate and measure that it so difficult to make reliable mathematical models to predict the behaviour of Earth's temperature in the future.

3

THE HISTORY OF THE ATMOSPHERE

STANLEY FELDMAN

DOGMA
We know that mankind causes global warming.

AS THE MOLTEN MASS that was to form our planet solidified, the gases from volcanic eruptions and constant asteroid bombardment came to determine the composition of its atmosphere. It is thought to have resulted in an atmosphere largely made up of water vapour, with around 15–20 per cent of carbon dioxide (planets such as Venus and Mars contain about 95 per cent CO_2 in their atmosphere). Other gases that were present, as minor players, included sulphur dioxide, carbon monoxide and nitrogen. Even before the advent of plant life and photosynthesis, the concentration of CO_2 was decreasing as more and more of it became converted to chalk and sand under the very high temperatures that were present as the fiery Earth started to cool. The greenhouse-gas effect of the huge amount of CO_2 in the atmosphere up to 50 million years ago did not prevent the molten mass of Earth from cooling, although it may have slowed the process.

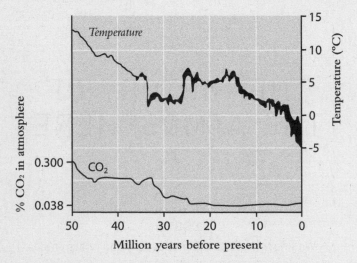

Figure 3.1: Deep sea temperature – in spite of very high levels of CO_2, Earth continued to cool. For the past 50 million years the level of CO_2 has remained historically low.

Although the first evidence of life can be traced back some 2–3 billion (thousand million) years, it was not until about 400–500 million years ago that the first recognisable animal life forms appeared. By this time it is calculated that the CO_2 concentration in the atmosphere had decreased from 15–20 per cent to 4–5 per cent. By the time the first marine organisms emerged on to the swampy land, many millions of years later, the CO_2 levels had fallen still further, to about 1.5 per cent in the atmosphere, many times higher than today.

In spite of the enormously high levels of CO_2 in the planet's atmosphere when it was born, the Earth cooled down. This is not surprising, since it was originally very much hotter than the surrounding atmosphere. There is evidence that the cooling continued, even when the surface became much colder, causing large parts of the planet to be covered with ice. It is believed that this occurred about 600 million years ago; it was the time of 'Snowball Earth', just before multi-cellular life forms appeared. The period of Snowball Earth was followed by several million years of warming.

The finding of fossil remains of gastropods in the trans-Antarctic mountains near the south pole is good evidence that the Antarctic was much warmer than today – indeed, warm enough to support animal life – about 15 million years ago. This means that the huge glaciation that formed the Antarctic must have occurred when the CO_2 levels in the atmosphere were over 0.5 to 1.0 per cent. Although no one can be certain of the CO_2 level at this time or the exact extent or duration of the glaciation, it is evident that, in the past, Earth has cooled to very low temperatures in spite of very much higher levels of CO_2 than is present today.

None of this is proof that CO_2 has not been important as a global greenhouse gas in the past or that it does not play a part in determining the present temperature of our planet. However, it does indicate that high levels of CO_2 in the atmosphere are not incompatible with global cooling. It must be borne in mind that many of the figures quoted are based on assumptions and calculations. This is why much of the argument about global warming has centred on what is happening now and how this relates to the period that we can measure with some accuracy, the past 1,000–2,000 years (see Figure 3.2).

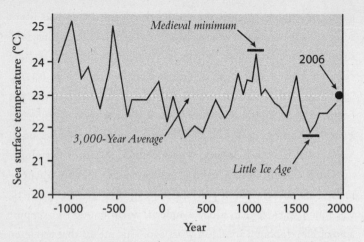

Figure 3.2: Temperatures on the surface of the Sargasso Sea over 3,000 years – evidence from isotope studies of marine organisms

Evidence from ice-core samples

Information about the levels of CO_2 and temperature dating back hundreds of thousands of years has come from cores drilled in the ice in the Antarctic. Bubbles of air trapped in the ice at the time it was formed give surprisingly reproducible measurements of the CO_2 levels. The temperature at those times is calculated from probes inserted into the ice which measured isotope ratios.*

They reveal that, for the past 500,000 years, the level of CO_2 has been roughly stable at around the present levels of less than 0.04 per cent. Over this period of 100,000 years the Earth has cooled

---

* Although these measurements have been fairly consistent and reproducible it is impossible to verify their absolute accuracy as there is no standard measurement with which they can be compared. It is inevitable that some CO_2, which is a soluble gas, will leach out of a bubble into ice, albeit at a very slow rate at these low temperatures. The magnitude of this effect must increase with time. What we can say with some certainty is that the levels of

slightly, although from time to time the temperatures have fluctuated through swings of 3–4°C.

In order to get some idea of what the conditions were like further back in time we have to make assumptions based on the changes that we know took place. Primitive cellular animal life evolved between 500 and 300 million years ago. As life developed and photosynthesis occurred, large amounts of the CO_2 were rapidly sequestrated from the atmosphere by plant and animal life and deposited as chalk, peat, shale, coal and oil. This produced a fall in CO_2 concentration from a high of 4.0 per cent to the 0.038 per cent found today. The white cliffs of Dover were made of their calcified carcasses, and they contributed to many of our hills and to the coral of tropical islands. Throughout this time, the Earth was cooling and, although the CO_2 levels were falling, they were many times higher than today.

The CO_2 produced today when we burn fossil fuels is merely returning into the atmosphere a minute part of the CO_2 sequestrated by plant and animal life over hundreds of thousands of years. It is not new CO_2 that we have produced. It is CO_2 that is being recycled. If we were to hold a huge bonfire and burn all the available fossil fuel, coal, oil and gas in the world at one go, it would raise the atmospheric CO_2 by only a small amount, to nowhere near the level it was 500 million years ago, when global cooling caused much of the Earth's surface to be covered with ice.

CO_2 recorded in these bubbles is the very lowest that would have been possible. They probably underestimate the actual values that were present all those years ago. It is probable that the actual CO_2 at the time they were formed would have been higher. Similar criticism can be levied against estimates of the temperature that pertained millions of years ago. These are not absolute measurements but are relative.

Predicting future global temperatures

It was almost entirely due to the apparent link between the temperature and CO_2 levels over the past few hundred thousand years, revealed by the analysis of ice cores drilled in the ice caps, that the present concern about the effect of human activity on the CO_2 levels in the atmosphere has been given some scientific credence (see Figure 3.3).

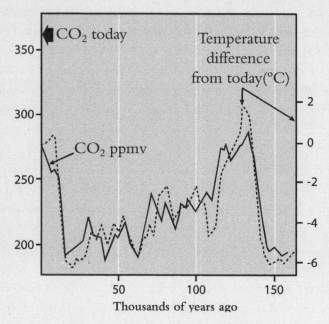

Figure 3.3: Ice-core date from Vostok, Antarctica
(after Raynaud et al., Science 259, 1993)

The indirect information obtained from the ice-core samples in the Antarctic give us a picture of the CO_2 levels and temperatures over the past 600,000 years. However, they can tell us only about the

CO_2 levels and temperatures locally; they do not tell us what was happening to the rest of the world. While it is likely that the CO_2 levels measured in one place, such as Mount Mauna Loa in Hawaii, will not differ significantly from that in the rest of the world, the same cannot be said for the temperature. It would be a nonsense to suggest that a period of hot summers in Manchester meant that there was also a period of excessive heat in Botswana. It is unlikely that the relative hot and cold spells found in the Antarctic ice-core samples accurately reflect what was happening in the rest of the world.

Although the results are largely reproducible, their absolute accuracy cannot be verified, as there are no standard measurements with which to compare them. However, we have direct and anecdotal historical evidence of the temperature changes in the past 1,000 years from various areas around the world (see Figure 3.4).

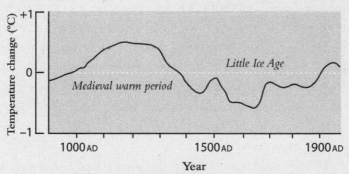

Figure 3.4: Over 1,000 years there is little change in
the temperature of northern Europe

Nevertheless, it is only in the past 50 years that advances in technology have allowed us to measure the change in CO_2 levels accurately and record temperatures continuously. In the past 15 years, thanks to the use of Earth-orbiting satellites, we are now able to obtain accurate, reproducible data on ocean levels, global temperatures and the behaviour of clouds.

Global-warming predictions

It is largely due to the gaps in our knowledge and the indirect nature of the information about events in the remote past that people have interpreted the same information in different ways. There is no doubt that some have seen political and personal advantage in filling in these gaps with the most absurd, scary guesses and projections; others have minimised their importance or ignored the potential problem; while some have used the information selectively in order to advance a preconceived hypothesis. The information has been misinterpreted and wildly exaggerated by some in order to make political points. Some pressure groups justify this as being necessary to alert the public to the perceived dangers of global warming. They echo St Paul, who in his letter to the Corinthians appeared to license such scare tactics when he wrote, 'If the trumpet give an uncertain sound who shall prepare himself for battle?' Indeed, Sir John Houghton, the first chairman of the International Panel on Climate Change (IPCC), is reported to have said, 'Unless we announce disasters no one will listen to us.'

There is no doubt that the evangelical nature and the scary exaggerations of many who claim the environmental banner – and their insistence that 'all reputable scientists agree with them' when, clearly, many do not (the Oregon petition had more than 33,000 signatures of scientists disagreeing with the theory of manmade global warming, more than 9,000 of whom have PhDs) – has done a grave disservice to science and made many question the basis of all the work linking anthropogenic CO_2 to global warming.

The most alarming suggestion of the doomsday environmentalists, that the oceans are rising at an alarming rate, has been proved wrong. Islands that were due to be submerged, according to their forecasts, are actually seeing the seas receding. Their prophecies have been shown to be wrong by information from orbiting satellites. Over the past eight or nine years the measurements from ERS1 and ERS2 (European remote sensing satellites) have demonstrated that there has been a rise in sea level of 0.5–0.1mm a year (within the margin of scientific error) and that the increase is not accelerating! Although there is some evidence that the levels of the oceans have risen slowly, possibly by about 5cm over the past 150 years, it is a process that started long before industrialisation produced an increase in atmospheric CO_2. So confident are the owners of property in the Maldives that the sea is actually receding that lavish seafront hotels are being built, yet the prophets of doom assure us that they will soon be under water. The island of Tuvalu in the Pacific, which was predicted to become submerged, has actually seen the sea level fall (it may be that the island is rising). The acceleration in the rate at which some glaciers in Greenland have been melting, adding to the ocean mass, started long before the motor car and industrialisation.

Not every drought or flood, hurricane or volcanic eruption, change in the climate or in the population of an animal species is due to global warming. There have been many such occurrences in the past and there is no evidence that their incidence is increasing (see Figure 3.5). Although the Arctic ice cap has probably been getting smaller since the Little Ice Age, most evidence points to an increase in the amount of ice in the Antarctic ice.

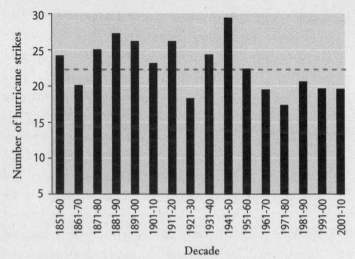

Figure 3.5: Number of hurricanes on the east coast of America over the past 150 years (National Weather Statistics)

Climate modelling

All the predictions of the future behaviour of the climate are produced from models that are based on past events. Most of the data used to construct these models comes from the same sources. If there are errors in this information it will be copied and exaggerated in all the models. Models of this kind have been likened to pop art, because, like the paintings of Andy Warhol, they project only the most obvious features on to a larger or longer picture. They are bound to reflect a personal interpretation of information rather than solid, verifiable evidence. They inevitably ignore some subtle details and they exaggerate others. They are limited by the impossibility of knowing every feature of a complex, multifaceted, ever-changing picture.

Many parameters, such as the exact contribution of water

vapour and of the clouds to global warming, are immeasurable or unknown. The practice, used in some of the scariest models, of labelling years as either warm or cold from the size of the rings in tree trunks has been shown to be so grossly inaccurate that the predictions based on them are considered fraudulent. The best model is only as good as the weakest information used in its construction. Many of the factors that might determine the global weather pattern, such as cosmic radiation and solar-magnetic effects, are omitted because they are uncertain or unpredictable.

In response to the criticism that instead of the predicted increase in global temperatures in the past nine years there has been a 0.4°C fall, and the failure of all the models to predict past events, such as 'The Little Ice Age'. A senior scientific adviser to the IPCC (International Panel on Climate Change), Dr Trenberth, admitted that the models were not intended to be predictions but were designed to cover a range of possibilities. They are meant only to be what-if scenarios. Unfortunately they have been treated as proven facts by the media, by politicians and by environmentalists who do not appreciate the level of uncertainty in their predictions.

Few weather stations equipped with the most sophisticated monitoring techniques would confidently predict the weather next year, let alone 50 years hence. The longer the timeframe of these projections, the greater will be the magnification of any initial error. It is little wonder that these models produce widely different predictions of future events. They often differ by a factor of 300 per cent. Bigger models do not necessarily produce more accurate predictions. Not all evidence is equally robust and, before decisions are made, based on these predictions, it is essential to appreciate their weakness. The possibility of bias and of a misinterpretation of the limited information on which they are based should be borne in mind.

Before we consider the evidence that underlies the various predictions that have been made about the consequences of the present and future levels of CO_2, it is necessary to ask the questions, 'Is the world getting hotter?' and 'If so, is it doing so at a historically unusual rate?' If these two questions give the convincing answer yes, then we must consider the evidence that this is due to an increase in atmospheric CO_2 caused by man.

Is the world getting warmer?

The answer to this question is that it all depends upon the length of the period studied.

If one takes a very long timescale – say 400 million years – then the answer is no: Earth is at least 10 degrees colder today. If one takes a shorter period – say 10,000 years – then the answer is yes: the world is slowly getting a little bit warmer. There is good evidence of what has been called the *medieval warm period* in northern Europe between 700 and 1000 CE, when temperatures soared. This was followed some 600–700 years later by the Little Ice Age when the Thames froze over and Bruegel painted his landscapes of the frozen Dutch countryside. It was probably during this period that there was a great expansion of the Arctic ice cap, extending it almost to the shores of Scotland. It caused the abandonment of the Viking settlements in Greenland, which before this time was covered with forests, and the closing of the Northwest Passage from the Atlantic to the Pacific by pack ice.

Coming nearer to our times, in the hundred years 1900 to 2000, northern Europe and America have warmed up by about 0.8°C and the whole world by about 0.6°C. This has occurred mainly as a result of warmer and shorter winters. However, between 1940 and 1975 there was reversal of this warming trend and Europe cooled by about 0.3–0.4°C. So marked was this cold period that predictions of an impending disastrous ice age abounded in the environmentally conscious 'green press'.

A word of caution about 'average temperatures'. Until recent times only maximum and minimum temperatures were recorded in many places. Often the places used as weather stations change over the years, and many are near town centres and may be affected by urban warming. During the breakup of the Soviet Union many stations were closed.

In the 25 years from 1975 the world warmed up by about 0.5°C, although this started to plateau off in 1998 (see Figure 3.6). In the past nine years, accurate monitoring of global temperature by orbiting satellites has shown that this period of global warming has come to an end. Instead of rising, there has been a fall in temperature of about 0.4°C.

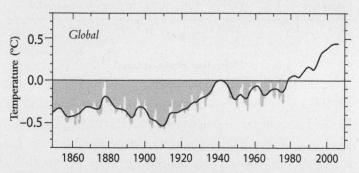

Figure 3.6: Global estimates of temperature change over 140 years (note the 0.6° rise from 1975 to 1998 that gave rise to the global-warming scare) (Calculation from Met Office Hadley Centre)

The net increase in temperatures, recorded over the past hundred years, appears to have been the result of fewer very cold nights, and only modest increases in the daytime temperatures. Table 3.1 demonstrates that the highest temperatures recorded were reached, in various parts of the world, over 50 years ago. Indeed, the number of days when the summer temperature has exceeded 32°C in the South of France has decreased markedly.

Continent	All-time high (°F)	Place	Date
Africa	136	El Azizia, Libya	13 September, 1922
North America	134	Death Valley, CA	10 July, 1913
Asia	129	Tirat Tsvi, Israel	22 June, 1942
Australia	128	Cloncurry, Queensland	16 January, 1889
Europe	122	Seville, Spain	4 August, 1881
South America	120	Rivadavia, Argentina	11 December, 1905
Oceania	108	Tuguegarao, Philippines	29 April, 1912
Antarctica	59	Vanda Station, Scott Coast	5 January, 1974

Table 3.1

How the scare started

It was the burst of warming following the cold spell that ended in 1975 that gave rise to the global-warming panic. Statistically, a hundred years of ups and downs of less than 1°C represents too little change over too short a period to know whether or not this is just another blip on the temperature chart or whether it represents the beginning of a new, potentially dangerous change in the world's weather. Much of the present anxiety is less to do with the actual increase in world temperature – after all, 0.5°C is less than the variation between town and countryside – but it is the rate of this increase in the 1980s and 1990s that alarmed the weather watchers and started the global-warming panic.

The question that faces us is this: does the recent increase in global temperature represent a departure from the recurring ups and downs of our weather and presage a permanent and increasingly rapid shift towards a warmer world? If so, can we or should we try to counter the effect?

It is like the problem of asking, 'How long is a piece of string?' that is at the core of our present confusion. There are those who paint a frightening picture of totally improbable 'tipping points' based on a speculative extrapolation of the most extreme predictions. They see the piece of string as being very short. They tell wild stories of cities under water, an ice-covered Britain and imminent global catastrophe. Every storm, drought or flood is proclaimed to be evidence of global warming.

Few of these claims stand up to scientific scrutiny. Climate changes have occurred from the beginning of time. There have been floods, drought and storms ever since the world began, they are recorded in the Bible and are present in the folklore of the first farmers who settled in Mesopotamia at the very beginning of civilisation.

Others take the view that the string is very long and point out that the world has seen the present changes over and over again during its history. They argue that a modest rise in temperature provides us with an extra source of energy at relatively little cost and, like Arrhenius and Callendar, they see the potential benefit to mankind. To them the string is very long. Unfortunately, good news does not sell newspapers or generate research grants, so one seldom hears their side of the story.

The majority take the view that the changes we are seeing are real and they may possibly be made more severe by industrialisation. They advocate caution, suggesting that we should do what we can to moderate the effects of any anthropogenic changes by anticipating any adverse effects that may ensue.

To claim that anyone, or any organisation knows whether the rise in global temperature in the past century is the result of human activity or just another blip in the history of our planet is tantamount to claiming that they have the answer to the conundrum: how long is a piece of string?

Carbon dioxide

Carbon dioxide, or CO_2, is a colourless odourless gas that is heavier than air. It dissolves in water, which is why the oceans of the world are such an important buffer against a rapid rise in CO_2. The amount that will dissolve in water depends upon its concentration in the atmosphere and the temperature of the water. As the temperature of the seas increases, their capacity to dissolve CO_2 decreases. The effect is similar to warming up a bottle of champagne: it causes the fizz of CO_2 to come out of the wine, causing it to froth.

When the world was formed, many billions of years ago, it came packaged with a finite content of minerals and gases, which have changed very little over all those years. What change there has been is the result of accretion from cosmic dust, asteroids and comets and as a result of radioactive transformation. These are believed to have produced a large addition to the total amount of water and minerals on the planet. As Earth cooled from the primitive molten mass, violent volcanic eruptions occurred during which vast extra amounts of carbon, in the form of carbon dioxide and carbon monoxide, were spewed into the atmosphere from the depths of the planet. Had nothing happened, the concentration of carbon dioxide in the atmosphere would have remained high, possibly reaching 20 per cent, approaching that found on other planets.

What has changed since that time is the chemical form in which the carbon is found. Over the ages, huge amounts of carbon dioxide have been taken out of the atmosphere and transformed into chalk and coral, into the hydrocarbons of coal, peat, methane and oil, into graphite, diamonds and charcoal. We are not producing any new carbon when we burn fossil fuels We are merely recycling atmospheric carbon dioxide in various chemical guises. The atmospheric CO_2 is constantly being taken up by plants in the process of photosynthesis. The plants are then eaten by animals and the carbon they contain is then given off when the animals metabolise fuels, such as sugar and fat, in order to produce energy. The energy used by a person on a bicycle is produced by burning a carbon-based fuel in the same way as a car burns petrol (but man is less efficient as a machine than an internal-combustion engine).

Carbon dioxide is not a pollutant (although the US Supreme Court ruled by five to four that it is). Like oxygen, it is part of our atmosphere and it is essential for life. Without CO_2, plant life would cease to exist and most species of animal life, including man, would die out. Ignoring the effects of industrial activity, the amount of CO_2 in the atmosphere is, in biological terms, a balance between animal and plant life. An increase in the amount of animal life, relative to plants, will increase the CO_2, while a relative increase in plant life will reduce it. The giant rainforests of the world take up some 30–40 per cent of carbon dioxide produced. Contrary to what has been portrayed in the media by pressure groups, 85 per cent of the Amazonian rain forests are intact and a further 12 per cent are in a state of 'recovering'. In the recovering state, the rapidly growing saplings are more efficient sinks of CO_2 than mature trees. Human activity and industrialisation contribute 2.5–3 per cent to the total CO_2 output. There is no doubt that this amount is increasing.

The concentration of CO_2 in the atmosphere is, in geological terms, historically low at 380 parts per million (0.038 per cent). As a result of this low concentration, it requires only very small additional amounts of CO_2 to cause a large *percentage* increase in its concentration. There is evidence to suggest that even at this tiny level it is about 0.008 per cent higher than it was a hundred years ago. Some of this increase is due to CO_2 that is given off by the oceans as their surface waters warm up, and a little comes from volcanoes, but most is attributed to human activity – the so-called anthropogenic CO_2. However, the actual total amount of CO_2 in the atmosphere is still tiny. It is estimated that, if no new CO_2 were added to the atmosphere, it would take only five years to reduce its level below what would sustain plant life; if plant life did adapt and continue at this low level of CO_2, all the gas would be used up in about seven or eight years. The CO_2 level found in our blood is essential to our wellbeing. A sudden reduction would cause many people to feel sick and faint and to develop headaches due to a constriction in the blood vessels supplying vital organs. We are physiologically far better adapted to dealing with a rise in CO_2 than with an acute reduction. This may be an evolutionary footprint reflecting the fact that animal evolution occurred during periods of higher, rather than lower, atmospheric CO_2.

To regard CO_2 as a pollutant is dangerous nonsense. It is less toxic than oxygen and it is present in the atmosphere at levels that are historically low. The problem that has to be faced is that we do not know with any degree of certainty what would happen if the levels were to be increased dramatically over the next century.

4

IS GLOBAL WARMING MANMADE?

STANLEY FELDMAN

DOGMA
Man is destroying the planet.

IF ONE ACCEPTS that the world has become warmer over the past century and this is a threat to our way of life, then it is necessary to examine the evidence that this is due to an increase in the amount of atmospheric CO_2 produced by man.

The only compelling evidence that CO_2 is linked to the recent rise in the Earth's temperature due to its greenhouse-gas effect comes from the studies of ice-core samples drilled in the ice caps. Although this may seem a somewhat inexact science, the picture they reveal is surprisingly reproducible. The temperatures at various depths can be measured and recorded, and the CO_2 content of the bubbles of trapped gas, while subject to a little variation, is also remarkably consistent. The dating of the samples is liable to small errors as the decay rate of the various isotopes measured can be influenced by outside events. Although these measurements cannot be considered definitive, our confidence in them is greatly enhanced because they have proved to be reproducible in different ice cores from varying sites made by different investigators. However, one should be cautious in interpreting these values as accurately reflecting the actual temperatures and CO_2 content of the atmosphere at the time the bubbles became trapped in the ice, since it is probable that some CO_2 will have leached out, over thousands of years, even into frozen ice.

The results of these investigations have been presented, in various forms, for up to 50 million years, but over 500,000 years their accuracy becomes increasingly questionable. It is the general close parallel, the casual association between the rise in temperature and the rise in CO_2 over the past 10,000 years, that has led to the present persuasion that the temperature change is driven by atmospheric CO_2. However, close examination of the samples from the Vostok Lake area of Antarctica over this period shows that in many instances the temperature increase occurred some 600 to 1000 years before the increase in CO_2.

Exactly the same association with temperature is also found when one examines the effect of the Milankovitch cycles (explained in Chapter 1). It is accepted that the periods of warm weather produced during these effects are due to the skewing of the Earth's trajectory around the sun so that it comes to be closer to the source of the sun's heat and is nothing to do with anthropogenic activity. It has been clearly demonstrated that, during each of the predicted Milankovitch warm spells, the CO_2 in the atmosphere increases. It is evident that a rise in CO_2 in this case is a consequence of the warming during these cycles and not the cause. We are at present coming to the end of a warm cycle that probably started at the end of the Little Ice Age some 400 years ago.

There is good evidence suggesting that the cause of some of the global warming is due to an increase in solar irradiance. There is a close correlation between solar activity and the atmospheric temperature in the Arctic between 1880 and 2000 (see Figure 4.1a). This association is much closer than that between temperature and CO_2 during the same period (see Figures 4.1a and 4.1b).

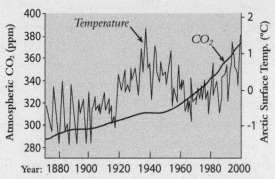

Figure 4.1a: Global warming and CO2 in the atmosphere over the past 120 years (source: W Soon)

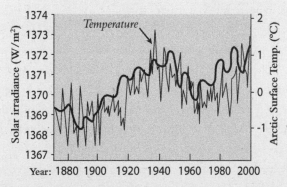

Figure 4.1b: Global warming and sunspot activity over the past 150 years (source: W Soon)

As the rate of CO_2 production has accelerated, it has led the International Panel on Climate Change (IPCC) in 2001 to predict, using mathematical models, that, without curbing the rate of CO_2 production, Earth's temperature will increase by between 1.4 and 4.2°C in the next hundred years. The assumptions made in these predictions include a continuously accelerating increase in population, no adaptive responses – such as a switch to alternative energy sources – and an accelerating growth in energy demand per capita. Using their figures and the presumption that it is CO_2 that is causing climate change, the mean of the most optimistic and the most pessimistic of the reasonable forecasts by the IPCC is that we may expect a rise in temperature by the beginning of 2100 of 2.8°C. This increase in temperature is about the same as the difference experienced between those living in London and those in the surrounding countryside and would render our climate similar to that of the South of France. Inevitably, it will be the coldest part of the day that will be most affected by this rise in temperature. The warmest days will not necessarily get much warmer. It will hardly cause the end of the world.

However, there are inherent problems in accepting the importance of the role of CO_2 in producing climate change. Apart from experiments carried out in the laboratory, the main evidence in support of the theory comes from one type of investigation – the ice-core samples. Associations of this kind between two events do not meet the criteria required for scientific proof. For there to be any confidence in the hypothesis, it is necessary to demonstrate either that lowering the CO_2 in the atmosphere produces cooling or raising it causes warming. Unfortunately, it is impossible to lower the CO_2

acutely but there is ample evidence that the world has cooled when the CO_2 in the atmosphere has been very much higher.

Medieval warming

It is generally accepted that the Earth warmed up significantly more during medieval times than during the 20th century and that it cooled down considerably during the Little Ice Age of the 16th and 17th centuries, without any anthropogenic event that would have affected the CO_2. Between 1940 and 1975, there was a significant fall in recorded temperatures at a time of intense industrial activity associated with the war and post-war reconstruction. In the last decade, the Meteorological Office Hadley Centre has confirmed that satellite records show that Earth has cooled by about 0.4°C, although the CO_2 level has increased by 0.004 per cent.

Although not absolute disproof, these well-documented events throw grave doubt on the validity of the hypothesis. Until there is proven confirmation of the relationship between CO_2 and the Earth's temperature, it remains an unproven theory.

If future technical developments show there to be a fault in the analysis of the ice-core samples that have been so readily accepted as proof of the role of CO_2 in global warming, then the whole scheme will be thrown into disrepute. So far, alternative methods of analysis of the age of the gases contained in the trapped bubble have supported the initial findings. The results appear to be reproducible. The problem with the ice-core investigations is that in some instances the rise in temperature found in these samples appears to precede the increase in CO_2. Many of the ice-core samples from around the Vostok Lake in eastern Antarctica showed that warming actually preceded the rise in CO_2, sometimes by 600 years. This effect had been noted in other ice-core samples but it is concealed by the compression of the timescale (x axis) in most presentations. Although various explanations of these findings, based on water–CO_2 feedback mechanisms (the Claussius–Clapeyron equation), have been offered by the CO_2 theorists, they are mathematically improbable and scientifically implausible. It remains absolutely impossible to explain how an increase in CO_2 could be the cause of a rise in temperature that occurred hundreds of years earlier. This finding, if proven beyond reasonable doubt in a single site, would constitute positive disproof of the theory that the increase in temperature is the result of the rise in CO_2. The rise in CO_2 found after a rise in temperature can be readily explained, as any increase in the temperature of the oceans would be expected to drive off dissolved CO_2. However, as they are a huge depository for energy this process occurs only slowly.

Equivocal evidence

There are many reasons why one cannot blindly accept the belief that anthropogenic CO_2 is the cause of global warming without reservation. The knowledge that the Earth has cooled in the past when the CO_2 levels in the atmosphere were many times higher than at present challenges the certainty with which the opinion that climate change is CO_2-driven is being presented. It suggests that those who tell us that the case is proven are wrong. The evidence is at best equivocal and our approach to the theory must be more open-minded.

There is little doubt that the world has warmed up over the past 50 years and that the rate of warming is a little greater than that seen previously, even if the extent is less than that which has occurred over other periods. Although it is possible to explain the rise in temperature by cyclical changes that occur naturally, it is the rate of change that has led to the speculation that some, if not all, of this is due to the extra CO_2 released by industrial activity. Certainly CO_2 is a greenhouse gas but the evidence that the extra CO_2 produced by industrialisation is the cause of the 0.6–0.8°C of global warming seen over the past 100 years, or that it presages a future calamity, is insufficient to be convincing.

Evidence for and against human activity as the cause of global warming

For

- A general correlation between temperature and CO_2 levels in ice-core samples going back 600 thousand years.
- An increase in human activity, CO_2 levels (0.008 per cent) and temperature (0.6–0.8°C) has occurred over the past hundred years.
- The demonstration that CO_2 is a potent greenhouse gas. It is not unreasonable, therefore, to believe that more CO_2 in the atmosphere will result in a reduction in the amount of the Earth's heat that is lost into space.

Against

- The correlation between CO_2 and temperature demonstrated in ice-core samples shows that in some instances the rise in temperature occurs *before* the increase in CO_2, often by about 600 to 1000 years.
- Earth was believed to be at its coldest 600 million years ago

(Snowball Earth) when the concentration of CO_2 in the atmosphere was up to 100 times higher than today.

- In the past 50 years, when the measurements are most reliable, the correlation between CO_2 levels and temperature is not very good. The correlation between temperature and solar activity is much better.

- There was a rise in temperature in the medieval period and a fall in the 16th and 17th centuries. Neither appears to have been associated with an abrupt change in CO_2 levels. The fall in temperature between 1945 and 1970 occurred at a time of intense industrial activity.

- The changes in temperatures have not been the same all over the world, although CO_2 levels in all areas are similar.

- 2007–08 has seen some unusually cold weather in various parts of the world with snow in Buenos Aires, Johannesburg, Athens and Shanghai. The coldest winter weather since 1941 was recorded in NE Australia; a very cold winter was recorded in the northeast of the USA; there was a cold spell in California in January 2008, which devastated the citrus crop.

- The CO_2 released by human activity comes from the carbon sequestrated from the atmosphere many years ago. It is not newly produced, merely recycled, and has therefore not added to the net level of the CO_2 on the planet.

- The amount of CO_2 in the atmosphere is minute compared with the amount of water vapour and droplets. Water is a much more potent greenhouse gas than CO_2.

Conclusion

It is reasonable to conclude that the world has warmed in the past hundred years and this has accelerated slightly between 1970 and 1998, after a cold spell in the 1940s and 1950s. This acceleration has now flattened off and there has been no significant warming since the El Niño peak of 1997; indeed, world temperatures have not increased since 1998. It is impossible to say whether this will remain within the limits of previous warm periods or whether it presages an exceptional period of global warming that will be a threat to mankind at some time in the future.

To say that it is proven that manmade CO_2 is the cause of present global warming is wrong. The evidence is, at best, equivocal. It is an unproven theory. The approximate coincidence between atmospheric temperature and CO_2 levels over many thousands of years, revealed from ice-core samples, is far from proof. A casual association falls far short of scientific evidence that is 'beyond reasonable doubt'. Yet it is the main evidence offered in favour of the theory. Prophecies of the future trends in global temperatures have been exclusively based on computer models. These are essentially self-confirming expressions of the original input dogma.

The great French scientist, Claude Bernard (1813–78) complained of

minds bound and cramped by their own theories and despisers of their fellows... They make poor observations because they choose among the results of their experiments only what suits their objective, neglecting whatever is unrelated to it and setting aside everything which might tend toward the idea they wish to combat.

43

Good science is about being sceptical; to accept a theory without good evidence because it is presented as a 'consensus view' is bad science.

A single proven ice-core result demonstrating, unequivocally, that the temperature rose before the CO_2 level would constitute absolute disproof and suffice to destroy the theory. There is good evidence suggesting that this has happened. Meanwhile, the lack of any significant global warming since 1998 suggests that the prophets of an accelerating disaster, of a hockey-stick-shaped increase in global temperatures, are wrong.

We just do not know what the overall effect of the CO_2, released by man's efforts, will have on the world's climate; it may take many years to obtain the necessary evidence. On present evidence, it does not appear to be a cause for panic. The risk of waiting until there is more certainty is much less than the scaremongers would have us believe. There is no Armageddon waiting around the corner. The end of the world is *not* nigh.

A final word

The evidence in the IPCC report (4AR WG I) is detailed and written in an authoritative style. It is presented as the bible of climate-change science. Readers should bear in mind that the working party that drew up the conclusions included those whose own opinions and papers are widely quoted in the compilation of the report. In effect, the report gives only their side of the story. The working party did ask some of those attending (many as representatives of governments that had approved the Kyoto Protocol) for comments on the draft Protocol. The most consistent comments were that it implied a degree of certainty that was not supported by evidence. Special criticism was made of the panel's reliance on predictions from mathematical models whose accuracy was disputed, even by the modellers.

The late Fred Seitz, a physicist at Oregon University Department of Science and Medicine, organised an online petition questioning the link between global warming and CO_2. The petition received 33,000 signatures from American scientists, of whom over 9,000 held PhD degrees.

In 2007, some four hundred climate scientists and astrophysicists from around the world, some of whom were on this or other IPCC panels (four times the number of those who drew up the IPCC report), produced a separate document (US Senate Report: '400 Prominent Scientists Dispute Man-Made Global Warming Claims') condemning the conclusions of the report as unproven, alarmist and wrong.

They believe that there is no evidence that the warming of the past hundred years is outside the parameter of natural temperature variability and they conclude that it is unlikely that there will be any significant warming driven by anthropogenic CO_2. They believe the scare story presented in the report is without scientific justification.

In spite of these massive petitions from scientists who do not believe that manmade CO_2 will cause a dramatic global warming, we are told that 'all scientists agree with the IPCC' and that hypothesis is proven. Clearly it is not.

Many scientists point out that there are very few institutions that can obtain funding for research that runs counter to the prevailing views of the IPCC, as a result one is left with an impression that there is no other story. It is difficult for those that question the anthropogenic global-catastrophe story to make their voice heard. The Compiler of Programmes for the BBC has said that she considers the case proven and that alternative views should not be given air time. The British peer Lord Lawson tells of the difficulty his agent had in finding an English publisher willing to promote a book that concludes that global-warming predictions are alarmist. The voice of respected scientists from outside their club of true believers is denigrated and they are frequently accused, by the green lobby, of being in the pay of the oil industry. The statement made by the chairman of the IPCC that there is a 'consensus in this science' is not based on fact: it results in insufficient attention being given to the views of the very large body, the silent majority, of dissenters. Good science is not served by promoting a false 'consensus'.

These chapters have been reviewed by five eminent scientists involved in energy science, climate physics and mathematical modelling. Two have asked for their names not to be revealed for fear of losing research funding or advancement.

I am grateful for the advice of H O Pritchard, professor of gas kinetics and combustion, York University, Toronto, Canada; Professor B Gray (emeritus), Sydney, Australia, at present combustion and scientific consultant, Turramurra, NSW, Australia; and Michael Arthur, geophysicist. They have all have given permission for their assistance to be recognised.

5

THE GLOBAL ICE CAPS

STANLEY FELDMAN

DOGMA

The ice caps are melting and will cause a 6-metre
rise in the level of the seas.

OF ALL THE SCARE STORIES about global warming, it is the
rise in sea levels and the fear of flooding of low-lying lands and
islands due to melting of the global ice cap that have caused most
alarm. Figures presented as likely scenarios have suggested a rise of
from 20cm (7.9in) (suggested by the IPCC in 2001) to 6 metres
(20ft) (suggested by Al Gore in his 2007 film, *An Inconvenient Truth*)
in the level of the oceans by the end of the century. Any rise in the
sea level will depend on global warming causing the melting of
huge amounts of land-based ice and dumping it into the sea at a
rate greater than that at which it evaporates. It is appropriate,
therefore, to question just what is happening to the global ice
caps, where most of the ice is found, and what is happening to

the level of the oceans, so as to determine whether there is a real cause for panic.

The Arctic ice cap

In August 2007, the world was stunned to learn that Russian submariners had succeeded in planting the Russian Federation flag on the seabed under the Arctic ice cap, claiming the landmass for Russia. The land is geographically an extension of the Siberian landmass, under the sea, giving their claim some validity. There was little doubt about the Russians' motives. They have no intention of settling the land, but are laying claim to the vast underground reserves of oil and gas believed to exist in the area. There are very good geological reasons why there should be oil under this part of the Arctic ice cap: it is an extension of the Siberian continent which has enormous oil deposits it is contiguous with the tundra shale deposits and the Alaskan oil fields that lie on the other side of the same Arctic ice cap.

The presence of oil under the Arctic demonstrates just how geologically new the ice cap is. Oil is a biofuel. It comes from the degradation of plant life. Originally it came from the carbon dioxide in the atmosphere and the water and minerals in the soil. If oil is there today, the land must have been covered by plant life millions of years ago. In the Antarctic, recent geological explorations have revealed the presence of fossils of primitive animal life, demonstrating that its mountains of ice are a maximum of 14–15 million years old.

The Arctic ice, which we are told is melting, is a geologically new phenomenon that covers much of what was once green land. It was lost in the last great Ice Age thousands of years ago. In winter it extends down to cover the green of Greenland. It extends from the North Pole down towards Siberia in the east and Canada in the west. It has frozen solid the Northwest Passage, which, according to history, was navigable in the 16th and 17th centuries. If there is a proven long-term diminution in the size of this gigantic iceberg, it will only be putting the environmental clock back to what it was very many years ago. It will hardly be 'destroying the planet'. But, then, is it actually melting?

There is no doubt that the size of the Arctic cap decreases in summer and extends again in winter. In April 2007, the decrease in size, as seen from orbiting satellites, was almost equal to the surface area of the USA. By October, the annual freeze was under way and the size of the ice cap was fully restored by November (see Figure 5.1). There was little if any overall melt. The icebergs that form during the spring as the ice cap melts provide environmentalists with photo opportunities. They have also provided a shipping hazard for hundreds of years; one of which caused the *Titanic* to sink.

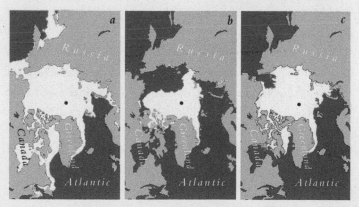

Figure 5.1: Sea ice extent in the Arctic (a) April 2007,
(b) August 2007, (c) November 2007 (source: Bremen, ERSP1 and 2)

The melting ice is not causing the extinction of the majority of the populations of polar bears. These animals are carnivorous and will follow the food supply, not the ice. They are extremely good swimmers and photos that purport to show them isolated on a raft of ice may have been artificially manipulated; they are unlikely to be a result of a natural occurrence. In spite of about 50–100 bears being shot each year, there has actually been a net increase from an estimated 500 in 1950 to 1,500–2,000 in recent times. Their survival depends on the availability of food, rather than ice. They live happily in the warmer climbs of Regent's Park Zoo in London. The threat to wildlife comes from man's encroachment on their environment, not from changes in temperature of less than one degree in a hundred years. Similarly, although some Antarctic penguin colonies, especially those near human bases, have decreased in numbers, the overall numbers of penguins are steady or increasing. The colony in the Coulman Islands and Cape Washington have an estimated 20,000 fledgling or breeding pairs. This has not prevented the Worldwide Fund for Nature (WWF), an aggressive environmental body, from claiming they face extinction.

The actual reduction in the size of the Arctic ice really depends less on absolute temperature changes than on the relative lengths of winter and summer and on the amount of rain that falls in the winter. The effect of global warming has tended to reduce the duration of the winters rather than to warm up the summers. Evidence from satellite photographs indicates that, in the past decade or so, ice has often been more rapidly lost in summer than it has been replaced in winter (a process reversed in 2005–08, when the ice cap increased in size).

The summer of 2008 was the coldest in Anchorage, Alaska, for

40 years. However, the absolute significance of any change in the Arctic temperature will be known only after 50–100 years. It is recorded that the temperature in this region was higher than today for a short time in 1998 and this had little if any long-term effect.

The temperature of the southern hemisphere, measured by NASA satellites, suggests that, over the past 100 years, the surface temperature has increased by about 0.05°C (see Figure 5.2).

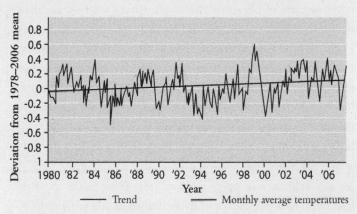

Figure 5.2: Temperatures in the southern hemisphere (source: NASA)

Nevertheless there does seem to be some decrease in the overall size of the Arctic ice cap in the past hundred years. As four-fifths of the ice is under water, this melt has not produced a noticeable rise in sea level. Even if it melted entirely – a very unlikely proposition because it is a gigantic iceberg – it will not cause more than a few millimetres' rise in sea level as all but its visible surface is already under water.

There would be a potential problem if cold water from the melting ice cap and Greenland pushed the Gulf Stream further south. It has been suggested that this may cool the seas around the

British Isles, causing colder, dryer weather. There have been dire predictions, from the environmentalists, that northern Europe will become a frozen wasteland. There is no evidence to support this view; in fact, the Gulf Stream is not getting weaker and it is getting warmer, not colder. Buoys in the Atlantic do not indicate any change in the strength or direction of the Gulf Stream.

There is rather more cause for concern about the land-based ice covering most of Greenland. Although neither the land nor the sea temperature has changed dramatically in the past 100 years and land-based ice is probably increasing slowly, there is evidence of a loss of ice from edges of this landmass and an increase in the rate at which the ice melt from glaciers is being added to the seas around the coast. This is believed to be the effect of water seeping through cracks in the glaciers, called *moulins*, and acting as a lubricating slide under large masses of ice, easing their passage downhill towards the sea.

An alternative explanation suggests that the landmass under the ice is being warmed by geophysical heat from below the Earth's thin crust, heat that cannot easily be dissipated due to the insulating effect of the overlying ice. It is known that Iceland sits on a tectonic ridge, which releases huge amounts of geothermal energy. It is a land of earthquakes and geysers. The significance of this 'glacier shortening' can be seen in perspective when one considers the evidence from isotopic studies of land-based organisms present in the Arctic seas. Although scientifically not as reliable as other studies, they do indicate that the increase in the rate at which glaciers have been melting started over 200 years ago, long before a rise in CO_2 levels (see Figure 5.3). The ultimate significance of this effect will depend upon the balance between new ice formation and the rate at which the ice is lost into the sea. Present evidence suggests that this is not a new phenomenon and that it is not specifically linked to anthropogenic global warming.

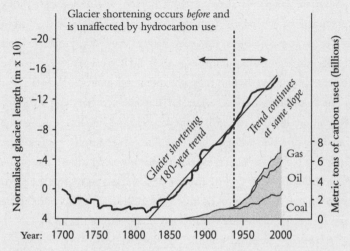

Figure 5.3: The rate of glacier shortening has been constant since about 1800 (after Robinson et al. (2007), Journal of American Physicians and Surgeons 12, p. 79)

What about the Antarctic?

Unlike the Arctic, the Antarctic ice is mainly heaped up on land. It is not possible to traverse the Antarctic by submarine as it is in the Arctic. Frozen sea surrounds the landmass of the Antarctic. The extent of the frozen sea depends on the seasons, as the migrating penguins know well. Like any iceberg, this ice ledge is vulnerable to an increase in sea temperature and there is some secondary evidence that the sea in this region is getting slightly warmer. Over the past hundred years a large finger-like iceberg projecting north into the Southern Ocean, the Larsen B ice shelf, has progressively eroded and partially collapsed. Like the Larsen B shelf, the Wilkins shelf, which collapsed in 2008, was a gigantic iceberg jutting out from the Antarctic peninsular; it was believed to have been less than a thousand years old. These ice shelves are not land-based and were always in danger of disintegrating during warmer years as they extended further north from the coldest region around the South Pole, and are subject to the strong currents eroding their underwater base. Like all icebergs, they did not cause any recordable change in sea level when they collapsed and melted.

It is difficult to be certain of the balance between new ice formation during the Antarctic winter and its loss during the warmer summer months, but satellite observations suggest that much of the Antarctic ice, especially that covering mountains in the eastern Antarctic, is getting more extensive. In this region the ice is now up to 3.2 kilometres (2 miles) thick. This is probably due more to an increase in rainfall than to any change in temperature.

The story is more complex in the western Antarctic, where coastal ice appears to be subject to the same sliding motion as in Greenland. It is believed that this is an effect of the formation of rivers and lakes under the ice. The loss of ice from these coastal regions is slightly greater than the formation of new surface ice further inland, resulting in a small reduction in the total amount of ice in this region. It is probable that this part of Antarctica is relatively new and that it has been subject to changes in the past. Evidence from fossils found in ice-bore samples in this region suggests that much of Western Antarctica is geologically less than half a million years old, compared with the 20–30-million-year history of the rest of the Antarctica.

The latest indications from the European satellites, ERS1 and ERS2, suggest that the total effect of the melting Arctic and Antarctic ice over the past nine to ten years has been to raise the ocean levels by 1.0mm a year. Although in geological terms, ten years is far too short a time in which to make any definitive judgement, it seems highly improbable that it will contribute more than a few centimetres to the height of the ocean by the end of the century.

Although more studies are needed before we can be certain as to what is happening, it would seem that the net effect of the changes in the Antarctic ice cap on the global sea level is minimal. There is absolutely no evidence that the world is likely to be flooded in the immediate future because of the melting of the Antarctic ice. At present, most scientists agree that the Antarctic is not warming significantly. There is little to support the view that a rise in manmade CO_2 is causing any of the changes seen at the ice caps, many of which started long before the present rise in CO_2 levels. While it is possible that some colonies of penguins may have been affected by man's intrusion into the frozen wilderness of the Antarctic, present evidence suggests the effect is small and that most colonies have increased in size.

Although changes are happening around the fringes of the ice caps, all the evidence suggests that they have been occurring for many hundreds of years. Until we have more information on the extent and rate of these changes, their long-term significance remains uncertain. Whatever is happening, it is unlikely it will cause the rise in sea levels predicted by the alarmist propaganda and the end-of-the-world prophets of doom.

6

THE ECONOMICS OF A WARMER EARTH

STANLEY FELDMAN

TO CONSIDER THE LIKELY economic impact of any future warming, we have to accept the theory that it is happening and that it is caused by man's activities. This allows us to set overall parameters of any probable temperature change and its possible economic consequences. Apart from the wildly alarmist and unlikely extremes of temperature predictions coming from mathematical models based on speculative suppositions, which have largely been discounted, we find most predictions are more or less in line with those suggested by the IPCC in 2001. (This is not altogether surprising, as they all base their predictive models on, more or less, the same data.) If the world continues warming at the same or slightly higher rate in the next hundred years as it has between 1975 and 1998, we can expect an average increase in the surface temperature of between 2.4 and 4.2°C. If the figure is based on the past hundred years, then the amount of warming to be expected will be less. This is unlikely to be spread evenly over the world. The predictions suggest that it is probable that some areas will see a change towards the top of this range, while others will see a less dramatic effect. The seasons are also likely to be affected. The winters will be shorter and the springs longer. It is predicted that we will see the nights becoming less cold rather than see a uniform increase in temperatures occurring at all times during the day.

The rise in sea level predicted by the IPCC 2001 of 20cm (8in) is possible only if there is a warming of the upper levels of the oceans and a melting of land-based ice. To achieve this degree of rise in the level of the sea, it would require the ice caps to melt at three to four times the rate seen over the past hundred years. Satellite readings of the change in size of these ice caps over the past nine years suggest that is not happening and, at the present rate, the sea level might rise by about 5–6cm (2–2.4in), not 20cm (8in), in the next hundred years. The IPCC figure would seem to be alarmist and at the very upper limits of any likely rise in sea levels, as it presupposes a huge increase on the rate at which the ice is melting and assumes a considerable increase in the rate of glacial melting. The evidence from satellites suggests that the loss of water by evaporation of the seas around the equator (even the Mediterranean loses water each year, which has to be replaced from the Atlantic) is, at present, more or less, compensating for any increase resulting from ice melt.

Effect on wind and rain

The effect of future global warming on rainfall and wind strengths is impossible to predict with any certainty. Many of these effects are cyclical and unpredictable and do not depend on the Earth's temperature. The problem can be addressed only as a generality. The amount of water on the Earth has hardly changed over billions of years. Most of it is either in the air, as water vapour and droplets, or in the sea. A warmer climate means more water will evaporate from the sea and be carried in the atmosphere as clouds. More clouds are likely to translate into a cooler Earth and an increase in rainfall. However, it is impossible to predict where and when the rain will fall, but extrapolating from past events suggests that it is likely that less rain will fall in areas around the equator and more in areas to the north and south of equatorial regions.

The higher the winter rainfall in the northern climes, the greater the formation of snow and ice (the opposite effect is demonstrated by Mount Kilimanjaro, whose ice cap has almost disappeared over the past hundred years, in spite of the intense cold at its summit, due to lack of precipitation). There has been a small increase in the amount of rainfall over the USA in the past hundred years. Even Australia, which suffers repeated serious droughts, has seen an actual increase in rainfall.

Those who confidently predict the drying out of central Africa may be right: it is after all a process that has been going on for many thousands of years. There is ample evidence that Chad and Botswana were once covered by vast lakes. The southern Mediterranean was much less arid. Even in Roman times it was the breadbasket of Europe. The low rainfall appears to have been caused by a change in wind pattern, which has produced a pattern of drought in the Sahara and heavy autumnal rain and floods occurring both further south and further north. It is certainly not due to human activity or anthropogenic CO_2 production. There is no evidence that a change in the atmospheric CO_2 will affect this process.

The behaviour of the main winds, the high-level jet streams and the ocean currents that circle the Earth causing the El Niño and Arctic and Antarctic winds is not easy to predict. They appear to be subject to cyclical variations. The Pacific conveyor currents and the El Niño have a cycle of about 18–20 years. There seems to be no good reason to believe they will alter their behaviour in response to a small rise in the Earth's temperature. The Gulf Stream, the Atlantic conveyer current, has not changed measurably over the past 50 years, in spite of predictions to the contrary. In general, winds are strongest when the temperature differentials between adjacent places are highest. Any equilibration of temperature between the tropics and the northern zones should, theoretically, reduce the incidence of tempests. There is no evidence that the amount of global warming over the past hundreds of years has increased the number or the severity of hurricanes in the Atlantic (see Figure 6.1).

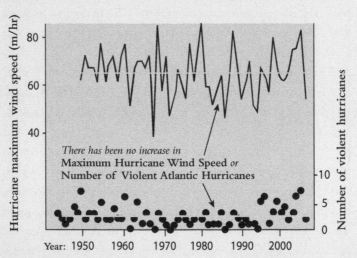

Figure 6.1: Number of hurricanes and maximum wind speeds on the Atlantic coast of the USA, 1944 to present day.

Economic consequences

The environment commentator Bjorn Lomborg has long argued that the economic impact of climate change would produce winners and losers, but, unless there was an increase in sea level above the highest of the ranges indicated, the economic benefits were likely to outweigh the losses. In a 2008 leading article on the proposals made in Bali by the environmental lobby, the UK's *Daily Telegraph* estimated that the effect, if fully implemented by all countries, would be to reduce the projected temperature rise at the end of the century by just over 0.13°C and the cost would be £5,000 billion. This imbalance in cost-effectiveness is in line with Lomborg's calculations for the likely effects of the Kyoto Protocol had it been fully implemented. Nigel Lawson, in his 2008 book *An Appeal to Reason* points out that, had the Kyoto Protocol been complied with, and all countries reduced their emissions of CO_2 to 5 per cent below the 1990 levels, the cost would have been financially astronomical and socially disastrous; the benefit would have been a reduction in any increase in global temperature of about 0.1°C by 2100.

On an overcrowded Earth with finite resources of fossil energy and water, it would seem that a climate change that promised to increase the supply of both should be looked upon as a blessing rather than a curse. A fall in temperature or a return to an ice age would be a calamity for mankind. It would use up the limited reserves of energy more rapidly; it would reduce the amount of rain and water available globally and expose vulnerable populations to the risk of deaths from hypothermia. A 2007 cold snap in Peru in 2007 caused 200 deaths from hypothermia. The colder weather would reduce the yield of crops significantly and those populations dependent on two crop yields per year would face starvation. We should stop complaining about global warming and embrace its potential, as Arrhenius and Callendar suggested, for what it is: a gift from nature. There is good evidence that a higher CO_2 level will increase plant yield and may help to offset the increasing demand for food as a result of the increase in world population.

Energy

The constraints imposed by the limits of accessible reserves of fossil fuel and a worldwide competition for energy are reflected in the rapidly increasing cost of petrol and gas. While huge reserves are present, they are either inaccessible or too expensive to utilise at the present prices. Most of the ready supply of fuel lies in the Middle East, a politically unstable area. The effect on the world economy of transferring huge reserves of money to such an unstable area is potentially dangerous. There is a level at which the production of more fossil fuel becomes so expensive that it is uneconomical to use. As a result, it is unlikely that the total consumption of these fuels as a source of energy will continue to escalate exponentially. The demand is likely to plateau off as it becomes increasingly unaffordable.

The present alternative of wind and wave power is expensive and of limited utility. Because the wind is intermittent and energy very expensive to store, it has to have substantial conventional energy-generation backup facilities. Nuclear power is cheaper to produce but it depends on the availability of uranium, itself a finite resource. Disposal of spent fuel and decommissioning costs are likely to be high. Unless nuclear fusion becomes a feasible proposition* we will run

*Prototype fusion plants are being built but are unlikely to contribute electricity to the national grid before 2018.

out of uranium for our reactors in about a hundred years. The effect of this will inevitably be to push up the cost of energy in the years to come and cause more uncertainty in the politics of the Middle East. The effects will be most severely felt by the poorest nations, who will be priced out of the energy market. It is likely to hold up their development and prosperity; it will make the poor poorer.

An increase in the sun's energy offers an opportunity to delay or offset this process. At present, solar energy is reliable and cost-efficient only in sunny climates. Provided we make the necessary investment in research technology, it could give us the means to harness this extra energy. This would be especially effective if large solar stations were established around the equator. The efficiency of these solar power stations can be greatly increased by the use of solar mirrors.

The alternative technology that might become economically attractive, if there is a large increase in the cost of fossil fuels, is the development of large-scale geothermal heat exchangers that harness subterranean heat. These techniques give us the potential to transform the sun's gift of extra heat into a portable form of energy, such as hydrogen.

Biofuels

At first sight, the idea of growing plants to trap the sun's energy to produce sugars and oils, which can then be readily converted to alcohol or diesel for use in cars and lorries, seems a virtuous circle, a win–win scheme. It is only doing what nature has done over the centuries to capture atmospheric CO_2 and, with the help of sunshine and rain, to convert it to fossil fuels such as petrol, gas and coal. The main disadvantages at present are the energy cost of doing what nature does for nothing: the conversion of sugar and fats into fuels. It has been estimated that it requires arable land the size of 30 football pitches to produce the biofuel necessary for a single flight between New York and London. The rush to biofuels has pushed up the price of corn and vegetable oils, which have had a knock-on effect on the price of animal feed. This has now reached a point where it is putting up the price of most foods and having a deleterious effect on the diet of many poor people in the developing world.

It must also be of concern to those worried about the amount of CO_2 that the fermentation process releases, although, on balance, this is far less than the amount taken in by the plants when they were growing. One can envisage a scheme whereby marginal land could be cultivated to grow crops that produce a high concentration of sugars, such as sugar beet, which takes large amounts of CO_2 out of the atmosphere as it grows. It is possible that, with a warmer climate and higher atmospheric CO_2 levels, genetic manipulation might produce faster-growing plants that would yield two crops a year on land now lying fallow. The extra amount of sugar grown could then be readily fermented to produce alcohol, a clean fuel.

Feeding the world

A warmer climate and an increase in CO_2 will be a boon for farming and agriculture in general. The probable advantage of such a combination was the basis of Arrhenius's predictions in 1905 that there would be golden years ahead for farming in Sweden. As recently as 1938 Guy Callendar predicted that a warmer Britain would lead to bumper harvests.

One can envisage returning to the Roman times, when vineyards were common in England. The importance of the wine industry in England in medieval times is reflected in place names such as Vine Street and Vintners Court in London. With less severe winters it will be possible to grow many crops that, because they are susceptible to the occasional frost, cannot be grown at present. Provided there is a supply of water available, the net effect for food production in the world will be positive. Contrary to the gloomy predictions of famine, food production is much higher in warmer areas of the world than in colder ones.

Keeping warm

Many more people die of the effects of cold than heat. We are physiologically better able to compensate for an increase in atmospheric temperature than for a very cold environment. Provided the humidity is low, man can survive at temperatures that would cook a steak, as a result of our ability to sweat. Immersion in ice-cold water usually kills in less than five minutes.

There should be little anxiety about the effect on health of a modest rise in temperature provided that there is access to adequate supplies of water and salt.

Floods

Anxiety about the effects of a rise in the sea level coupled with extra rainfall has led to the fear that flooding will be a major consequence of a warmer Earth. If the oceans do rise by up to 20cm (8in) in the next century, as suggested by the IPCC, then it will potentially cause a loss to those living in low-lying flood plains and islands, if no precautions are taken. Their predictions were based on the belief that melting of the land-based ice caps and their glaciers was accelerating. Although some melting has increased, the evidence suggests this is to a large extent balanced by new ice formation. Even if the sea levels rise, evidence from the past suggests that, given the slow timeframe of any likely change, it is probable that vulnerable populations will take measures to minimise the risk.

There is plenty of evidence that the seas have been both higher and lower in the past. Towns that were on the sea in biblical times, such as Ephesus, are now 15km inland, whereas in other areas Roman, Greek and Indian cities can be found submerged below shallow seas. Most of these changes have occurred as a result of earthquakes, volcanic eruptions or shifts of tectonic plates that are still taking place. Few have been the result of the small increase in the levels of the oceans. Flood defences are expensive but less so than the cost of trying to hold back any rise in sea level, like a latter-day Canute, by reducing industrial activity.

Higher sea levels and tidal surges, when wind and tide converge, will increase the risk of coastal flooding. Since most people enjoy a 'sea view' and are willing to pay extra for the privilege, it has become attractive to build in areas that are likely to become vulnerable to any increase in sea level, causing a headache for the insurance industry.

The potential cost of building sea defences, river barriers and drainage systems will be high, although economists differ in their assessment of these costs. There is little doubt, however, that the greatest risk, in terms of the financial loss compared with their income, would be felt by those living in the coastal areas in the developing world if, as the IPCC predict, the seas levels suddenly start to rise.

Fortunately, unlike the alarmist forecasts of the extremists, the evidence from satellites suggests that the timescale for any significant increase in ocean level is likely to be long enough for protective and adaptive precautions to be made over the course of many years. After all, the Dutch managed to build sea defences in the 16th century that have stood the test of time.

It has been suggested that anthropogenic global warming is the cause of excessive rainfall, although it has also been blamed for droughts. Although there does seem to be some slow changes in the pattern of rainfall over the past century, all the evidence suggests that these changes are the result of alterations in air currents in the upper reaches of the atmosphere. In their evidence to the Pitt Review, reporting on the floods in the UK in 2007–08, the Flood Prevention Society presented evidence that the annual average rainfall has not altered over the past hundred years in the UK. It just happens that some years are wetter than others. The assertion that the floods were due to global warming is not based on any known science but it distracts attention from the underlying failure of the authorities to maintain riverbanks and flood-drainage systems.

Sustainability

The need to carefully husband finite resources, such as fossil fuels and food, and to use them effectively is clear. To this end, energy conservation and the use of alternative sources of energy are to be applauded. Too often this sensible objective is confused with that of reducing CO_2 emissions in order to limit global warming. Both the threat of competition for finite resources and any effect of an increase in the amount of CO_2 in the atmosphere originate from the same source: the rapid increase in the animal life, especially human life, on the planet. The amount of CO_2 in the atmosphere is fundamentally a balance between that produced by animal life and that taken up by plants. As the amount of animal life on Earth increases, the amount of CO_2 produced goes up. Animal life increasingly encroaches on the remaining uncultivated, wild land, reducing the amount of forest and plant life, leading to a reduction in its ability to take up CO_2.

At present there is pressure on the food supply due to a rapid increase in the world population and an increasing demand as a result of the newfound affluence of developing countries such as China and India. An increase in CO_2 and a rise in mean temperature should help to offset this problem by producing an increase in agricultural yields. With the possible exception of the more arid counties around the equator, global warming should be a boon in helping to meet the increasing demand for food. We should welcome the gift of any extra energy in the form of warmth; we should be glad that the CO_2 in the environment is not falling. A low atmospheric level of CO_2 and a cooler world would make it impossible to grow the food necessary to feed the world.

There are many, including the environmental agencies, pushing their own global-warming agenda, politicians who see political advantage in playing on the fears of a global catastrophe, and UN bodies, such as the Committee on Sustainability (chaired by Zimbabwe), who claim that, unless we reduce consumption, stop our dependence on fossil fuels and return to what they claim is a sustainable rural lifestyle, the planet is doomed. There are worrying portents of what might happen if we were to follow their advice. The effect of recent rises in the cost of fuel and food and the credit crunch has produced an insight into the likely effect of such a reduction in consumption. It has led to food riots in developing nations, fuel protests in countries of Africa, Asia and South America, strikes in Europe, panic meetings of international financial bodies, government intervention in some of the hardest-hit developing countries and pledges of emergency aid from the UN. It is the poorest people in the least well-off countries that are hardest hit. To try to turn the clock back so that we all live in a so-called sustainable state is dangerous nonsense. It is arrogant of those who enjoy the benefits of an industrial society to deny those benefits to many of the less fortunate in the world.

Cost

Various economists have tried to grapple with an analysis of the likely cost of an increase in global temperature and a rise in sea levels. Any such analysis must start with assumptions that are themselves suspect: that the increase in temperature is driven by the rise in anthropogenic CO_2 and that it will continue to escalate, without respite, over the next hundred years. We just do not know if this is reasonable or true. The various models of the likely effects have too many unknowns to be taken as reliable, although, given the above reservations, they all point in the same direction. Of the studies reported, most have concluded that the net effect will be an increase in cost. Although the amount varies, most accept a figure of around 1 per cent of gross domestic product over a hundred years. Since the GPD is expected to rise by significantly more than 1 per cent over this period, the cost will merely mean the world will get richer slightly more slowly. The damage sustained by the increase in the risk of flooding will depend on the future use of development land in coastal areas and flood plains and upon the investment in infrastructure, but all agree it will be a significant part of this cost. In this respect the UK government-inspired Stern report * is much more pessimistic than others suggesting a figure nearer 10 per cent of GDP. Lord Stern reached this alarming prediction by constructing a model based on all the most pessimistic prognostications of the most costly effects of global warming and ignoring the positive beneficial ones.

All the predictions of the cost of global warming suggest that, even in the very unlikely event that the rise in ocean levels exceeds that forecast, the cost will be many times less than that of attempting to meet the Kyoto Protocol target.

The injunction of the environmentalists to reduce consumption in order to lessen the CO_2 output of industry is extremely unlikely to be heeded throughout the world. If it did happen it would have a disastrous effect. While the prospect of a period of negative growth in the developed world would cause massive unemployment and unrest, it would also frustrate the ambitions of the developing world and would cause a further fall in its standard of living and introduce the probability of mass migrations. The rise in global fuel and food costs experienced in the first half of 2008 and the effect of the credit crunch in the second half of 2008 have demonstrated that any move to further restrict growth, as demanded by the environmental fundamentalists, will affect the poor more than the rich and is likely to cause political unrest and riots in the poorer countries.

* Lord Stern was the economic adviser to the Blair government whose scientific adviser was Professor David King, who was among the more alarmist in his predictions on the effect of global warming when Stern was commissioned by the government. Tony Blair had already decided that global warming was 'the greatest threat to mankind' and staked his political credibility on the need to reduce UK CO_2 emissions.

Rio, Kyoto and Bali

Looking back at the repeated swings in temperature that have occurred over the past hundred thousand years, it is evident that the present warm period is within the range of changes that have been experienced previously. In the not-too-distant past, sabre-toothed tigers roamed the land and malaria was endemic in the UK. Periods of ice-cold weather were interspersed with years of balmy climes. Whether global warming will prove to be a benefit or a disaster will depend upon the rate with which any further change occurs and our ability to meet new challenges. One thing is clear: a cooling world would impose a much greater threat to our way of life and to social stability than one that is gently getting warmer.

In spite of the lack of any evidence that anything unusual is happening to our weather, the professional protestors have been busy. They preach doom and disaster; it has become a religious belief (indeed, not to believe it has become heresy in the eyes of certain church leaders). As a result of their fervour and the effectiveness of their propaganda, the spotlight of public concern has focused on the environmental impact of industrialisation. By raising an irrational fear, based on an unproven theory, they have bullied politicians and supranational bodies into demanding precautionary changes in the way we live.

It is evident that the bulk of these changes are aimed at protecting the developed nations and will ultimately be paid for by lower standards of living in the developing world. By compounding two totally different topics they managed to turn a very real concern – that we may be running out of energy resources and food – into a debate about the influence of CO_2 on climate change. Unperturbed by the fact that only 30 years earlier the same environmental lobby was scaring us about the coming ice age and the need to change our way to avoid inevitable doom, these self-proclaimed guardians of our planet now proclaim the end of the world from anthropogenic global warming.

The very pressure groups that campaign to stop CO_2 production obstruct the building of nuclear power plants, the only really feasible way of producing a reliable source of energy without producing CO_2. They recognise the need to produce more food but oppose genetically modified crops that can help solve the problem. They advocate organic farming, which is 40 per cent less efficient than conventional methods.

When, in 1938, Guy Callendar predicted global warming, he pointed out its huge benefits; when Arrhenius suggested Sweden would get warmer due to a rise in CO_2, he saw it leading to bumper harvests. The propaganda of the present-day soothsayers predicts death and destruction in order to capture the attention of the world. Their case is irrational. Fear is no substitute for good science.

The result of the pressure from the propagandists was a meeting in Rio in 1992 under the auspices of the UN. The United Nations Framework Convention on Climate Change (UNFCC) was dominated by politically active environmentalists with the stated intention of 'stabilising greenhouse gas emissions at a level that would prevent anthropogenic changes in global climate'. It started from the supposition that the theory of anthropogenic global warming was a fact. In so doing, they prejudged the science by taking for granted that any increase in the temperature of the planet was due to man. They sidelined scientists who disagreed with their preconceived notion. The Rio Protocol called for huge cuts in energy use by the industrialised nations. Although signed by

President Bill Clinton, it was never sent for ratification to Congress, who had voted unanimously not to take any measures that would interfere with the US economy, unless every country was prepared to share the burden.

The Rio congress demonstrates a total lack of common sense. It made King Canute look like a genius. It was followed by the Kyoto meeting in 2001 and the so-called Kyoto Protocol, based on the UNFCC. The Protocol called upon the developed world to cut back its CO_2 emissions, to 5 per cent below the 1990 levels. There was never any hope that this ambitious target would be met. Many large polluters, including China, India and the USA, refused to sign the Protocol. Russia, whose industry had collapsed following the break-up of the Soviet Union, initially refused to sign the Protocol. But it realised that the establishment of CO_2 trading, by means of which heavy polluters could buy absolution for any transgression of their limit, would benefit it greatly, and eventually agreed to sign the Protocol.

In fact, by 2007, only two European countries had come near to meeting their Kyoto targets. China alone was increasing its CO_2 output each month by more than the total UK emissions in a year. In 2007, a meeting was held in Bali to try to reconcile the implacable objection of the USA to penal targets, with the avowed intentions of the green movement to reduce energy consumption. It was an attempt to achieve the impossible, especially if it was to achieve its aim without hurting the nascent industries of the developing world, many of which are energy-inefficient.

In all these meetings, it has been stated as an act of faith, by the prophets of this new religion, that without a reduction in CO_2 production the planet is doomed. It is a belief that ignores the wide fluctuations in the levels of CO_2 and in the temperature of the Earth that have occurred in the past. It arrogantly suggests that the paltry 3 per cent of the 0.038 per cent of CO_2 in the atmosphere that is produced by man's activity is controlling the temperature of the world. The ice-core samples, on which the CO_2 theory of global warming is based, demonstrate that levels of this gas in the atmosphere and the surface temperature of the Earth have been going up and down for hundreds of thousands of years, long before man started making fire, wore clothing or made flint daggers.

The past variations in CO_2 in the atmosphere have very little to do with man's activities. They are presumed to be the result of natural events, animal activity, volcanic eruptions, forest fires and changes in the sea temperature. Ninety-seven per cent of all the CO_2 comes from non-anthropogenic sources. If we were to go back to living in caves and walking everywhere, we would still generate 97 per cent of the CO_2 now entering the atmosphere. It is produced by all animals when they breathe. It is inevitable that, with an increasing human and animal population, more and more CO_2 will be produced. This effect is much more significant than the meagre saving possible by extreme measures to reduce domestic and industrial CO_2 production. Even if this were achieved, the lessons from the history of the atmosphere demonstrate that climate change can still occur. Cutting back on manmade CO_2 will not alter the primary importance of the principal factors affecting the temperature of the planet: the sun, the clouds and water vapour and droplets.

The assumption that it is politically possible to reduce consumption and return to a Stone Age culture is absurd. All we can hope to do is to limit the rate of future increases in CO_2 against a background of an increasing world population and a desire by more and more people to have access to the benefits of civilisation. Even if it were possible to stabilise CO_2, the eruption of a good-sized volcano, a widespread forest fire or the loss of a significant wooded area would more than negate the effectiveness of the measures proposed.

CO_2 trading

The trading of CO_2 allowances is an attempt to ration limited resources in a way that permits flexibility necessary for industry. The concept is to give each industry a quota that represents a part of the country's total target CO_2 allowance. Should the need arise to use more energy and produce more CO_2, then that company, or country, can buy the allowance or quota from a source that does not need so much energy, either due to reduced demand or increased energy efficiency. Although a sound concept in theory, it is open to abuse and it is doubtful whether it can be fairly applied. It is hugely expensive with CO_2 brokers and the energy industry creaming off large profits.

Among the problems is how to allocate targets for emissions. Should it be on a per capita basis or on a per capita GDP basis? Who is to set the targets and who will police them? The USA produces many more goods than India. Should this mean it is allowed to produce more CO_2 per person? If China uses twice as much energy to produce a car as in Europe, should its allowance be based on the number of cars? What about countries rich in non-polluting sources of energy? Should they suffer the same limits on growth? How do you stop rich countries and global companies shifting their heavy, energy-using, environmentally polluting industries to poor countries to make use of their exemptions from CO_2 trading?

Should some of the financial burden be shifted to those countries who produce the CO_2-releasing fuels in the first place?

The concept of CO_2 quotas has led to the trading of CO_2 'offsets'. By purchasing a 'green credit', a CO_2-producing activity can be granted absolution. For example, a CO_2-offset company in the Mediterranean offers the owners of a mammoth gas-guzzling super-yacht a chance to offset the CO_2 they produce by buying shares in a wind farm. With such a purchase the CO_2 produced by the yacht is apparently no longer deemed to be a cause of global warming!

There are many problems in applying the system worldwide. Inevitably, it will put up the cost of living, an effect likely to be felt most acutely by the poor. It will frustrate people's natural desire to live a fuller life with access to modern technology, drugs, travel and food. It will slow the industrial development of poorer countries, restraining their aspirations. It will transfer large amounts of money from large manufacturing countries, especially the USA and Europe, to less industrialised countries such as Russia, China and the Middle East. It is likely to have an acutely depressing effect on the economies of the major countries in the world.

At the end of this costly exercise, trillions of pounds will be added to the cost of living, unemployment will rise to a level that may cause economic disruption and, even if it met the most optimistic forecast and assumptions, it is calculated it would reduce global warming by less than 0.2°C in a hundred years.

2

THE EPIDEMIC OF OBESITY AND MATTERS OF HEALTH

7

GOOD FOODS, BAD FOODS

VINCENT MARKS & STANLEY FELDMAN

DOGMA

Junk foods are bad. They cause obesity and heart disease.
Salt can kill you.
There are superfoods that will prevent you getting
old or dying.

Healthy lifestyles

WE ARE ENCOURAGED to follow a healthy lifestyle, consume a healthy diet and exercise regularly and often, in the belief it will keep us fit. What started as encouragement has now turned into a threat to deny treatment in the National Health Service to those who fail to live according to the government's 'prescribed healthy lifestyle' and to reward those who do. We are to be tested, inspected, measured and weighed so as to make sure we are conforming to the dictates of this regime and are not abusing our bodies in a manner that could, in the opinion of the policymakers, cause disease or, more particularly, cost the health service more money.

It seems that the policymakers have rediscovered the wheel. The belief that prevention is better than cure was behind Beveridge's assumption that a National Health Service would prevent ill-health and therefore the need for medical treatment. As a result, it would ultimately save money. It turned out to be a very buckled wheel. It does not take a genius to work out that it is unlikely to work, as preventing one cause of death will merely lay one open to another, possibly more expensive, way of dying.

Life, after all, is a fatal disease. Few reasonable people would deliberately set out to choose an unhealthy lifestyle, although a few do ignore the early signs of a potentially curable disease, and, in spite of good evidence of the risks involved, maintain their right to smoke or drink excessively. Most people would prefer to be active and healthy; few would choose to be unhealthy and chair-bound. The trouble is that, although we know some of the things that shorten our life expectancy and make us vulnerable to disease, there is relatively little that we can do that will positively result in our living longer and healthier lives.

We have been told we must give up smoking, as it causes cancer and heart disease, eat fewer eggs, drink less milk and cream and ration our intake of cheese and butter in the belief that the cholesterol they contain causes heart attacks. Never eat any egg or egg product that has not been hard-boiled to kill the salmonella they were once thought to nurture. Avoid chilled foods, as they are a listeria risk. Bacon, sausages and hamburgers are junk foods high in dangerous chemicals and saturated fats, which are bad for our hearts. Sugar, fizzy drinks and sweets cause dental decay and are not only bad for your health but they make you fat.

We must limit our diet to one portion of red meat, ham or smoked food a week for fear their nitrosamine content will cause cancer. We must avoid salt and stop eating any salty food because it causes high blood pressure and strokes. We must not drink more than five glasses of wine or beer a week and it must be avoided completely during pregnancy.

To achieve a healthy lifestyle one must exercise regularly and often, go on a two-mile walk every day, swim for 30 minutes, or go to a gym and work out. Never smoke or inhale other people's smoke. Never go out in the sun without first using sunblock. Never have unprotected sex (presumably an exemption is given for those couples wishing to have a child).

You must measure your blood pressure often, have your blood examined for kidney failure and diabetes. Your weight should be measured frequently to make sure you maintain a body-mass index (BMI) under 25. Drink two litres of water a day to wash out the toxins in your blood. Have an ultrasound examination of your aorta (if you are over 65), have cervical smears once a year (unless you are under 24), have any itchy dark spot removed surgically. You must have breast screening every two years. You must be vaccinated, immunised and have an influenza jab every year and eat five pieces of fruit or vegetable every day to ward off cancer and old age.

In fact, none of these stratagems can assure long life or good health. Many are useless; others, while probably harmless, are of unproven value.

Eggs seldom contain salmonella, even if some chickens do. Cholesterol in the diet does not cause atheroma (fatty deposits) in your arteries. There is probably little difference between the effect of saturated and unsaturated fats. In those with normal kidney function salt does not cause high blood pressure. Those with a BMI between 25 and 32 live as long as or longer than those with a lower BMI. Drinking extra water is useless, it merely causes more trips to the lavatory, unless you are in a hot climate or involved in strenuous physical work. Avoiding the sun causes vitamin D deficiency; suntan is nature's sun blocker, although sunburn is to be avoided. Heterosexual spread of AIDS is possible but unusual in the UK. Breast screening generally picks up the more benign slow-growing tumours rather than the really dangerous ones. If everyone who was found to have an aortic bulge had surgery the surgical death rate would almost certainly outnumber the lives saved. Most dark itchy spots are benign, moderate sunlight is good for you and vigorous exercise is no better than moderate exercise, especially in older age groups.

Although we do know that much of the advice, especially on foods, is flawed and has caused a needless interference in people's life, some of the suggestions have stood the test of scientific scrutiny. Smoking does kill, although there is less convincing evidence about secondary smoking. A BMI of over 32 does reduce your life expectancy. Cervical screening does reduce the number of cases of cancer; breast screening in those at risk is probably worthwhile. Reducing high blood pressure by taking drugs has reduced the incidence of strokes, although salt restriction is useless. Immunisation has reduced the incidence of childhood diseases. Moderate exercise and activity is associated with a better quality of life as well as longevity. Breathing cleaner air has reduced the incidence of emphysema and bronchitis (although, surprisingly, it appears to be casually associated with a higher incidence of asthma). However, many of those living to a healthy old age have often ignored all of this advice. Indeed, one lady of 103 recently said she attributed her longevity to eating bread and dripping every day, while another elderly man claimed his long life was due to often eating ham, bacon and pickled onions.

Bad foods?

Possibly the greatest interest centres on the notion that we can remain fit and avoid disease by cutting out certain 'bad foods' from our diets. It is variously claimed that 35–50 per cent of all cancers are caused by the food we eat. However, it is undeniable that not eating any food at all causes a 100 per cent death rate! What then is the truth about bad foods, the so-called junk foods or fast foods? Are they doing us harm? Are they, as the TV chef and food polemicist Jamie Oliver claims, clogging the arteries of children, causing obesity, constipation and bad behaviour? Should they be singled out for yet another government health warning?

Junk food

The term 'junk food', as it is applied to hamburgers, sausages, pizzas and other fast food, clearly has nothing to do with nutrition. It is a pejorative oxymoron. All nutrition is food and no food is junk.

I can remember how, when my children were young, I would buy them toys beautifully fashioned out of polished wood with a satisfyingly smooth feel, only for them to be left on the shelf in favour of coloured bits of plastic. Later, they failed to share my enthusiasm for well-crafted clothes made from hand-woven wool and preferred linen jackets that always looked as though they needed ironing, sweatshirts, trousers of denim, and chinos. This paternalistic arrogance about what is good for others is demonstrated by the fashionistas of food, the celebrity chefs and self-styled nutritionists in their attitude to what they term 'junk food' or 'fast food'. Their disdain for mass-marketed, cheap products makes them ignorantly claim that they are bad, while high-fashion, expensive and exotic individually prepared food is pronounced to be good. Thousands of healthy adults have proved them wrong

The term 'junk food', like the commonly associated alternative 'fast food', has no meaning. If a food has a nutritional content, it is clearly not 'junk'. The term 'fast food' is equally silly. Some foods require more cooking and preparation than others; some take longer to eat than others; but few would lump all of the ones that are quick to cook or eat as bad 'fast foods'. Although despised by the culinary elite, the readily available hamburgers, sausages and pizzas have provided good nutritional value for many low-income families who in previous days could afford only low-protein, high-carbohydrate, high-fat meals such as bread and dripping or chip butties.

The big breakthrough that these convenience meals have made is in bringing down the cost. This has been achieved largely as a result of a rapid turnover in outlets that do not encourage diners to linger. Although it is possible that it is this availability of food at an affordable price that has stoked the so-called obesity epidemic, this is not the fault of the food but rather that of those who unthinkingly gorge themselves.

The antagonism to hamburgers and sausages is especially irrational. Most countries have a national dish based on minced or processed meat. Meat balls are used in many guises in the Middle East, chopped meat on a bed of onions is a national dish in Bosnia and the Balkans, moussaka is a national dish in Greece, chopped meat is used in lasagne and bolognaise sauce in Italy, and minced lamb in shepherd's pie in the UK. The terrines and pâtés of France and Belgium contain processed chopped meat.

All these foods use minced or chopped meat in a manner similar to that of hamburgers. The protein in the meat used in these dishes has exactly the same chemical composition as in steak or a joint from the same animal. Subject both to chemical analysis and the same amino acids will be found in the minced meat as in other meat from that animal.

In his history of the hamburger, Josh Ozersky pointed out that in 1930 a burger manufacturer demonstrated that a student survived for 13 weeks eating only hamburgers and drinking only water. According to the University of Minnesota, who controlled the experiment, he remained in good health throughout. How different from the effect that Morgan Spurlock claimed to have suffered from eating McDonald's hamburgers for a month. Unfortunately, neither study has great scientific validity.

Defeated on the basis of hamburgers' composition, the self-styled nutritionists point to a possible higher fat content, although this is not always true. The critics who denigrate the hamburger and the sausage because of their potentially high fat content also advocate cheese, either alone or as a topping or on a pasta. The fat in cheese is very similar to that in the meat product, and both are animal fats. Nevertheless, there are some who are taken in by the term 'junk food' and they want them banned.

Following ingestion, a beefsteak is minced by the teeth before it is digested in the gut. It enters the bloodstream as its constituent amino acids and fat. These building blocks will be identical irrespective of the part of the animal from which it originated. A rump steak, a fillet steak, a rib eye or a hamburger all end up as the same amino acids in the blood. The much disparaged Turkey Twizzler, made of recovered turkey meat, provides the same amino acids as turkey breast. Corned beef, now an unfashionable meat product, is no less nutritious than any other beef, although, like Turkey Twizzlers, it is a reclaimed meat product. Various cuts of meat may have different flavours, different fat contents and textures but this is not necessarily reflected in their nutritional value.

To denigrate a meat product because of either its fat or its carbohydrate content is equally silly. After all, meat is commonly eaten with potatoes or bread and followed by cheese or an oily dressed salad. Provided the standard claimed for the product is met, additional fat or carbohydrate is acceptable. Indeed, low-fat hamburgers and sausages are fairly dry and tasteless. It is often the fat that provides the flavour.

It is about time we stopped using meaningless terms such as 'junk food' and 'fast food' and concentrated on the contribution these convenience meals can make to a balanced diet. Watching the approach of some French schools to teaching children healthy eating illustrates the difference between those in a country that takes healthy eating seriously and one under pressure from celebrity chefs who do not understand the physiology of nutrition and try to make children eat food that adults enjoy. In France, hamburgers are on the menu, as are cheese and sausages. Children are not force-fed unappetising pastas and green vegetables. The emphasis is on eating a variety of foods with a balance of familiar ingredients and of eating only moderate-sized portions.

Just as there are no foods that will prolong a healthy life, there are no junk foods that will kill us. It is the totality of the diet, the mixture of the types of food and the variety of tastes that are important. Provided a mixed diet of carbohydrate, fat and protein is eaten, together with sufficient essential vitamins and trace substance, there is no evidence that any particular food product should be avoided.

Salt

Before the advent of drugs that lower the blood pressure, the only recourse available to physicians treating patients with high blood pressure was to restrict their intake of salt. This seldom worked, as a low-salt diet of boiled fish, vegetables and rice is unpalatable and the effect on the blood pressure is negligible. Nevertheless, the myth has persisted that high blood pressure is caused by eating too much salt and the government has bullied the food industry into reducing the salt in foods and tells us to cut our intake to a measly 6 grams a day (this would make most cheeses unacceptable).

Salt is an essential food. Without it we would die. Animals in the wild deprived of salt develop a craving that causes them to seek out 'salt licks'. The evolution of the animal species has developed in oceans, an environment high in salt (although the sea was less saline than today). They have developed an efficient mechanism, with checks and balances, to rid their bodies of any excess salt through the kidneys. Land-based mammals control their body temperature by sweating and panting. Sweating is impossible without sufficient salt. Strenuous exercise in a person depleted of salt causes overheating and death. It is probable that some of the sudden deaths in healthy marathon runners are due to inability to lose heat due to lack of salt. Elderly people die during heat waves if they are short of salt and it may contribute to deep-vein thrombosis, which occurs in some people during long-distance travel. Even in cold climates sweating is essential, as the body produces a large amount of heat when it metabolises food and produces energy.

A large intake of salt causes the body to retain fluid and may cause a short-term rise in blood volume. In most people this is readily accommodated in the very distensible venous system. In a few older people with rigid vascular systems it may cause a temporary rise in blood pressure. The increase in blood volume causes the kidneys to excrete more salt in the urine over the following 6–24 hours. If too little salt is eaten the blood volume is reduced, causing a release of hormones in the body, which cause the blood vessels to contract and, in some patients, the blood pressure to rise.

There have been numerous studies on the effect of giving extra salt and of diets that are low in salt. The results of 11 of the scientifically most credible studies involving 3,500 patients were analysed by the internationally recognised Cochrane Collaboration. They found a negligible overall effect of salt intake on blood pressure. The Parliamentary Office of Science and Technology (POST) 2003 report quoted the result of one such investigation, the Intersalt study:

> In the largest study (Intersalt study) which collected data on salt intake and blood pressure from over 10,000 adults from 32 countries, no statistically significant overall association was found between salt intake and blood pressure.

Salt intake among the Japanese is about twice that in Europe, yet they have a longer life expectancy and no particular problem with high blood pressure.

Sick patients who have to be fed intravenously, often for weeks on end, have a fluid regime with the equivalent of 9 grams of salt every day.

The British government has caved in to the anti-salt zealots in its advice to reduce salt intake. There is very little, if any, truly scientific evidence that it does any good. There may have been some point in it if there were no other way of treating high blood pressure, but there are anti-hypertensive drugs available that are effective and cheap. They have relatively few side effects. Since their introduction, the incidence of strokes, while still too high, has been dramatically reduced. There is good evidence that salt restriction can do harm, especially in the elderly or those indulging in vigorous exercise. The government advice on salt intake should come with a health warning!

Can food be bad for you?

There is little doubt that some ingredients taken in food or as food additives can cause harm. An excessive alcohol intake can cause liver disease and mental deterioration. The regular use of potent spices such as hot peppers and curries may be associated with the high incidence of gastric and oesophageal cancers found in Southeast Asia, and betel-nut chewing, common on the Indian Subcontinent, like tobacco chewing in the Americas, is undoubtedly a cause of cancer of the mouth and tongue. Foods contaminated with high quantities of heavy metals, especially mercury and cadmium, can cause disease, but one would have to eat a lot of them. It is probable that other exotic herbs and spices that are used in some countries to add flavour or colour to their meals may cause harm if used too frequently. However, the body can usually cope with small, occasional quantities of all but the most potent poisons.

In most cases, it is not an individual food that reduces life expectancy but an inadequate or unbalanced diet. Starvation causes organ degeneration, and that frequently leads to a premature death. Poverty has always been associated with a lower life expectancy and much of the blame for this lies with the inadequacy of the diet. Pockets of extreme poverty, such as occurs in parts of Africa, China, India and Russia, have resulted in a failure of the life expectation to increase significantly, or actually decrease in these countries. The diet of most poor people consists of too much carbohydrate, which is cheap, too little protein and fat and a lack of vital trace elements and vitamins. Inadequate diets are a much more serious problem than the consumption of any particular bad food. Supplementing these inadequate diets by readily available so-called junk food, with its cheap protein and fat content, should be applauded rather than condemned. It results in the addition of nutritional supplements that make up for the deficiencies in many poor diets. It is absurdly smug of celebrity chefs in rich countries to rail against healthy, cheap, mass-produced foods when it is these very foods that prevent many poorer people suffering from the ill effects of an inadequate, life-shortening diet.

The food we eat

I (SF) have just eaten a fishcake produced by a leading food-processing company. I chose to eat it because I liked its taste and texture. Like most of the food we eat, we do not choose it because of its content of any particular ingredient. Food is more than a refuelling exercise. It was pointed out on the wrapper of this fishcake that it contained only 20 per cent fish, much of the rest being carbohydrate. Does this make it less desirable, or less good for me, than a fishcake containing 30 per cent fish? As far as I am concerned, it is the taste and appearance that matter rather than the relative percentage of this or that ingredient. In the Western world it is unlikely that anyone on a normal diet would be any better off if the fishcake had a higher concentration of fish protein, especially if this meant it did not taste as good. Much of the extra protein will end up in the urine as urea; the rest will become body fuel. Meats sold because they contain little fat are no better or worse than those with some fat. After all, the fat in meat is similar to that found in cheese and it is the fat that gives flavour to meat. Yet low-fat meat is highly promoted and sells at a premium.

The importance of recognising that it is personal taste that determines the type of food we eat was brought home by the introduction of so-called 'healthy meals' in schools. Children will not eat food they do not like, even if they are told, often erroneously, that it is good for them. The take-up of school lunches fell by 30–40 per cent when celebrity chefs started dictating what the children should eat. Providing a balanced diet is more important than any consideration of the particular form in which it is eaten; that will always reflect personal preferences.

8

AN EPIDEMIC OF OBESITY?

VINCENT MARKS

DOGMA
There is an epidemic of obesity.

AN EPIDEMIC is defined by *The Oxford English Dictionary* as 'a sudden, widespread occurrence of something undesirable'. It is used in its proper sense in relation to influenza, for example, as described in Chapter 14. It is wrong to describe the increasing prevalence of obesity in communities throughout the world as an epidemic. The increase in corpulence that we are witnessing has been going on for at least a hundred years and probably as long as food has been readily available and affordable.

In the past, 'obesity' was a word used only by the medical profession; now it is mentioned every day in newspapers, magazines, radio and television programmes along with 'overweight', as though they were synonymous and clearly defined. Neither is true. Governments issue lengthy statements and make laws intended to prevent obesity – seemingly with little effect – and 'bosses hire diet police to help staff shed weight'. What, then, is obesity? Is it a self-inflicted condition or is it an illness? Why are we said to be suffering from an epidemic of it when clearly we are not?

Body shape

Until recent times, being thin was always associated with ill health and looked upon, quite rightly, as a consequence of sickness. Plumpness, on the other hand, was a sign of good health and something to be admired and striven for. The paintings by Rubens and other portrayers of human form illustrate this to perfection. Excessive weight – fatness – on the other hand was recognised as a physical handicap and something to be avoided if at all possible. The most grossly overweight were looked upon as freaks. They were almost invariably young people who topped the scales at 200kg kilos (31 stone) or more and died before the age of 40.

It was a layman, William Banting, a moderately well-off London undertaker, who, in the middle of the 19th century, and on the advice of his doctor, popularised the idea of dieting – recommending a diet (it was similar to the Atkins diet) to reduce his own weight. Banting's excessive weight was modest: at the age of 66 he weighed 91kg (14 stone), giving him what we would now call a body-mass index (BMI) of 33.5 – little more than it had been for much of his adult life. It nevertheless reduced the quality of his life, though clearly not its quantity, as he lived to be 81, considerably longer than average for the time. The fact that Banting's book went into four editions and sold more than 60,000 copies shows that corpulence was far from uncommon in Victorian England and probably in other industrialised countries as well.

Corpulence

Corpulence – or obesity to give it the name by which it is now universally known – is specifically the result of excessive accumulation of fat or adipose tissue. It is not the same as being big or weighing more than average. Obesity develops when the energy *content* of the food taken in exceeds energy *expenditure*. It is maintained when the balance between energy intake and expenditure is stabilised, as invariably occurs in most cases. It is almost always due to a combination of genetic disorders, the exact nature of which are only now beginning to be uncovered. The majority of really fat people – just like those of more modest dimensions – reach a weight plateau which they maintain for most of their adult lives. The questions we have to ask ourselves are these: are fat people at risk due to their corpulence (or otherwise inconvenienced like William Banting)? Does it shorten their lives? Is it likely to kill them at an earlier age than their parents? Before these questions can be answered we need to define what we mean by obesity and what is an ideal weight.

Is there an 'ideal weight'?

This is a crucial question. Banting introduced the concept of an ideal weight from a health point of view in his little book, which he called a pamphlet. He based it on information given to him by a doctor, involved in the life-insurance business, who recognised that the length of survival depends upon a number of factors, including both bodyweight and height. This relationship is now usually expressed as body-mass index (BMI) which, as a determinant of longevity, is far from perfect. The figures he arrived at for men of between 1.55 and 1.86 metres tall, about 5 feet to just under 6 feet (a 6-foot man would have been considered a health risk!), after conversion to BMI, produced an ideal BMI of between 23 and 24.5. This was remarkably prescient for a layman. Banting's recommendations were confined to men and took no account of age; women were not considered. The ideal weight for a women was catered for by the fashion industry. It fostered the idea of plumpness during the 19th and early 20th centuries but, in the late 20th century to the present time, changed to equate slimness with wealth and success. In recent times, plumpness has been identified with dowdiness and sloth – exactly the opposite of its previous image.

The ideal woman, according to the fashion industry, is one who is visibly thin and can show off her highly priced clothes to advantage. The quest for feminine slimness originally had nothing to do with health and everything to do with perception. Nevertheless, it has led to the belief that to be thin is to be healthy. The famous remark attributed to the Duchess of Windsor, probably incorrectly, that 'you can never be too rich or too thin' is still widely, but erroneously, held to be true. Adherence to its philosophy has been held by many psychologists and sociologists to be responsible for the enormous increase in cases of anorexia nervosa that have occurred in recent times. Actually, to be too thin is as great a risk to health as being too fat.

Ideal weight tables

No one has ever denied that extreme fatness, like extreme thinness, limits life expectancy – but at what level of 'fatness' does the risk to health increase above 'normal'. Banting's table was a start at putting the subject on the agenda. It was, however, based on a sample of only 2,468 healthy men and does not appear to have been adopted. No serious attempt to define ideal bodyweight for the general population was made until the lifetime survival figures, generated by Metropolitan Life Assurance Company, were published. They were based on data collected from hundreds of thousands of men and women whose lives they had insured. People were described as overweight when they exceeded ideal weight for their 'build' – a mythical concept introduced to take into account people's various shapes and dimensions. 'Obesity' was a term confined to the medical profession to describe people who were so overweight that they actually complained about it to their doctors – in other words, they became patients.

Advances in medicine also confirmed what many doctors had always believed: that obesity predisposed people to a number of chronic illnesses, of which diabetes type 2 and hypertensive heart disease are probably the most important. In these instances the association can be demonstrated to be a cause of the increase in both morbidity and mortality from these diseases and one that can be lessened by weight reduction.

Measuring obesity

Various ways have been devised for measuring obesity. From a medical point of view obesity has less to do with a person's weight than with the percentage of it that is made up of fat (adipose tissue) and how much of muscle, bone and other tissues. There are numerous ways of doing this but they are all complicated and do not necessarily provide consistent results nor do their implications for longevity necessarily agree with one another. To complicate matters, we now realise that it is not just the quantity of fat but how it is distributed that is important. In the medical profession this has led to a simplistic but useful differentiation of obesity that occurs in people who accumulate fat mainly above the waist – especially in their abdominal cavity (apples) and those who accumulated it predominantly below the waist (hips and thighs), fancifully identified with pears.

The BMI was devised in the 19th century by a Belgian epidemiologist, Adolph Quetelet. It relates weight to height through a formula in which weight in kilograms is divided by height in metres squared. It correlates reasonably well in statistical studies with degrees of fatness assessed by other methods. But it is not suitable as an assessment in everyone, as it says nothing about the proportion of bodyweight that is fat – the real test of obesity. Nevertheless, more than a hundred years after it was described, the BMI was enthusiastically adopted by epidemiologists, who could use the single number it produces to enter into their computers, write lots of grant applications and publicity-attracting papers about the growing prevalence of obesity, without advancing the understanding of the condition a jot.

Quite when, and on what basis, it was decided that a BMI of 20–25 is 'normal' and one of 25–30 is 'overweight' and consequently unhealthy is far from clear. Banting's data did not support these figures and there has been no subsequent evidence to substantiate them. Nevertheless, it is so ensconced in the folklore of the subject that even those whose own research brings it into question treat it as gospel. There is good evidence that the conventional BMI is no more reliable as an indicator of obesity in women than one based on weight divided simply by height. In spite of all the clamour about childhood obesity, there are no reliable data for what constitutes health-threatening obesity in children.

Analysis of insurance company statistics was the original basis on which 'ideal bodyweights were established' (Metropolitan Life

Assurance tables). Reanalysis of the data collected on 4,200,000 policies taken out between 1954 and 1972 revealed that, as people get older, ideal body BMI, from the perspective of longevity of life, increases progressively from about 20 for young adults to around 28 for those over 70. Plumpness in late middle and old age, especially in women, is much less of a health risk than it is in the young and early middle-aged. Indeed, statistically, it is a distinct advantage for longevity to be plump in old age.

Undoubtedly, the increasing age of our population is a factor in the rising prevalence of plumpness revealed by epidemiological studies, which have been wrongly interpreted to imply that community health is deteriorating rather than improving. Several large prospective studies emanating from the National Institutes of Health in the USA, and involving tens of thousands of healthy people, have confirmed that there is nothing to differentiate between the life span of people with BMIs between 20 and 30 at the time of their entry into the study. These findings have been consistently confirmed in other countries. No studies that contradict this conclusion have withstood critical scrutiny but this has not prevented the obesity zealots from continuing to label healthy people with a BMI of between 25–30 as overweight and those with a BMI greater than 30 as obese.

Like all conditions from which mankind suffers, obesity is the result of the interplay between nature, in the shape of genetic and antenatal factors, and nurture, principally in the shape of the availability of food. While this may seem obvious, it has not always been accepted. It is not difficult to see why. To the casual observer, (most) fat people eat no more than thin ones – indeed, they often seem to be remarkably abstemious – while some thin people appear to be bottomless pits into which food could be shovelled with seemingly little effect. There is no doubt that this perception is wrong over the long term.

Statistically, fat people both expend and consume more calories than thin people, although the overlap between them is enormous. This is mainly due to differences in resting metabolism – the amount of energy required just to keep the body warm – and the levels of physical activity required in order to carry out normal living. It is quite easy to show mathematically that quite subtle differences in food intake or energy expenditure could, over a period of many years, produce profound changes in body shape. For example, taking in the energy contained in just one spoonful of butter more than expended every day would, after a year, theoretically cause a 2kg (4.4 pounds) gain in weight.

This simplistic approach to the causes of obesity belies the complexity of the situation. What is truly remarkable is how most people manage to maintain more or less the same weight once they reach adulthood without any conscious effort to control what they eat. It is as if they possessed a 'bodystat' analogous to a thermostat in a refrigerator. The mechanics of this bodystat are still being unravelled by biochemists, physiologists, psychologists, sociologists and epidemiologists throughout the world.

The role of genetics in the process of getting fat has been known to farmers, veterinarians and experimentalists for over a hundred years, but was established in human beings less than 50 years ago. It began with studies of obesity in identical and non-identical twins, where the effect of environment, especially access to food, could be controlled. It is now recognised that there are more than 40 genes that are linked, in some way or other, to the propensity to become obese. Although simple single-gene types of obesity occur, they are extremely rare.

Obesity in childhood

If a strict and usable definition of obesity in adults is acceptable though imprecise, in children it is almost impossible. Once upon a time, plumpness in childhood was considered an achievement; now it has become a stigma and something that politicians and some paediatricians go to great lengths to try to prevent. There is, however, no evidence that this is either necessary or achievable in a free society. Genuine obesity in children – however it is defined – is more likely to be due to a genetic or glandular disorder than to wilful gluttony and it is relatively rare. When it occurs, it requires investigation and treatment.

Plumpness in adolescents, especially girls, is often described as 'puppy fat' and is very common. Evidence linking it to adult obesity exists but it is poor; most plump children do not become obese adults and most obese adults were not plump as children. The link is stronger for the adolescent than for younger children, in whom the link between plumpness and adult obesity is so weak as to have little predictive value. This is not to deny the value of trying to prevent children from becoming too fat, if only because of the psychosocial damage it does them through ill-informed prejudice. Unfortunately, this prejudice is being made worse by the authorities, who demonise children who are overweight.

Adverse health effects of obesity

No one disputes the real hazard of gross obesity, which can arbitrarily be defined as a BMI greater than 35 – and which, in spite of what one reads in the media, is still quite rare. Lesser degrees of 'obesity', arbitrarily defined as a BMI of 30–35, are also associated with a slight reduction in life expectancy that is age-dependent. This disadvantage in longevity becomes less as one gets older and may even be positively advantageous in people over 65. Failure to accept this demonstrable medical fact is responsible for a lot of misery among the increasingly large number of elderly people.

There are some conditions, most notably Type 2 diabetes and hypertension, that undoubtedly are causally associated with even moderate plumpness. There are also a number of other illnesses, notably some cancers (especially of the bowel, breast and ovaries) whose incidence is modestly increased in people with BMIs above the norm for their age. In general, however, current evidence suggests that plumpness in middle age does not decrease life expectancy and in old age it improves it. Life expectancy is probably the best single indicator of 'good health', but must be distinguished from fitness (fitness for what?) and feelings of 'positive wellness' that have no established link to longevity but contribute to the quality of life. For example, such conditions as osteo-arthritis are more likely to impair the quality of life of a heavy person than that of a thinner one without reducing their life expectancy.

Gross (true) obesity

Gross obesity has been observed in some people since the earliest of times but has always been looked upon as a medical curiosity, although in certain primitive societies it was once considered a royal attribute. Gross obesity also occurs in several inbred strains of rodent and a study of these has provided insight into some of the physiological mechanisms in the maintenance of bodyweight and its disturbances.

In one such inbred strain of mice the specific gene responsible for obesity was called ob. Only mice inheriting a copy of the gene from *each* parent develop a condition that results in their depositing so much fat that they weigh four to five times as much as their siblings who have inherited one or no copy of the gene.

Mice inheriting the double dose of gene eat ravenously until they are so fat they cannot move to get to their food. However, when treated by restricting the amount of food available, they become just as slim and as active as their littermates, only to revert to gross obesity and immobility when food is supplied ad lib.

We now know that these fat mice lack the ability to make a hormone called leptin, which has the ability, among its many properties, to suppress appetite. Within the past 15 years, a similar condition to that of the double-ob mouse has been identified in human beings but it is exceedingly rare and accounts for only a minuscule proportion of grossly obese children. It can be treated specifically with leptin in exactly the same way as children with diabetes can be treated with insulin.

It came as a great disappointment when it emerged that the overwhelming majority of obese people have more rather than less of the average amount of leptin in their blood. This is because it is fat itself that is the source of leptin. Leptin acts principally on the brain, where it forms part of the bodystat mechanism. Fat people also have more insulin in their blood due to a similar mechanism: as obese people are less sensitive to its actions than slim people, they need more of it to maintain their bodily functions and their pancreas obliges by producing it.

In nature, an animal's appetite stimulates it to seek food. This declines as its weight rises from the accumulation of fat when food is readily available. The same applies to most human beings. If there is no limitation on the availability of food – as exists for most of the people living in the developed world – a person's weight may increase to such an extent that it imposes a threat to their health. Eventually, the higher concentrations of leptin produced by the increased amount of adipose tissue in their body suppress their appetite to such an extent that they eat only sufficient to meet their (increased) daily requirements for energy. Consequently, their weight stabilises at a new level.

Leptin is, however, only part of the story. The past 20 years has seen the discovery of many different hormones, all of which play a part in the regulation of body fat. We are still a long way from understanding exactly how they work and interact with one another. Some, such as leptin, are mainly concerned with regulating appetite, while others, such as GIP, a hormone produced in the gut when food is eaten, are more concerned with regulating how food energy is disposed of within the body, either by storage or by (inefficiently) burning it as waste.

Hormones and obesity

When I first started researching in obesity, I often met patients who vowed that they didn't eat excessively and that it was all due to their 'glands'. The idea that hormones played a part in controlling bodyweight began with the discovery that patients with thyrotoxicosis, caused by an overactive thyroid gland, often developed ravenous appetites yet lost weight. Conversely, those with an underactive thyroid often gain weight, though they very rarely become genuinely obese.

It was only after we became able to measure hormones in the blood that they have once again entered the arena as an important factor in the genesis of obesity. Thyroid hormones play no part, whereas insulin, much better known for its role in diabetes, is crucial to the development and maintenance of obesity. Hormones produced in the intestine in response to food were also suspected of playing a role in obesity, but have only recently been shown to do so. It was, however, only with the discovery of leptin and other hormones that are concerned with the control of appetite that interest in the role of hormones in the causation of obesity has really taken off.

Unfortunately, we still have a long way to go in understanding how these more recently discovered hormones work to control bodyweight, although they are already being used to develop potential treatments. Their investigation has helped explain why some surgical treatments for obesity work much better than others. It has been shown that in those who also have Type 2 diabetes surgery may effect at least a temporary cure. Genetically engineered animals, rendered insensitive by genetic manipulation to GIP, do not become as obese with overfeeding as their normal littermates and seem to suffer no other disadvantage. This supports earlier work that suggested that GIP is one of the factors that led people to become obese. It may serve a survival purpose in situations where food supply changes from glut to famine.

Contrary to popular belief, it is extremely difficult to become clinically obese by voluntarily eating excessively. This was established by experiments performed on healthy young volunteer prisoners who were encouraged to eat as much food of their choice as they wanted. They became fat but not obese. Similarly, laboratory rodents become fat – though nowhere near to the same extent as in the genetic varieties – when fed a cafeteria diet that contains more sweet, fat food than the boring feed they would normally get.

These experiments show that mere access to unlimited supplies of food is not enough for the average rodent or person to become pathologically obese: something more is required. It clearly has something to do with appetite and the ability to overcome the feeling of satiety that most people experience when they have eaten sufficient for their physical requirements. Leptin is one component but there are many more. Some increase appetite, others reduce it. One is a recently discovered hormone called PYY, which is produced in the intestine in response to food. It, like leptin, works on the brain to suppress appetite. Under experimental conditions, PYY enables people to resist the temptation to eat excessively. It is currently being pursued as a potential treatment for obesity. Another gut hormone affecting appetite is one called GLP-1, which stimulates the pancreas to secrete insulin and is used for the treatment of Type 2 diabetes. It has the important side effect of reducing bodyweight. How useful these and many other newly discovered hormones and their antagonists will become in the treatment of genuine obesity remains an open question, but understanding how they work in the body will help put paid to the ludicrous idea that true obesity is all due to gluttony.

Harm due to obesity

Plumpness or overweight in adults is not life-limiting, especially for those past middle age. It appears to be as much a natural physiological state as what is arbitrarily called 'normal weight'. It becomes harmful only when it exacerbates an underlying pathology.

Nevertheless, for many people, especially women, plumpness is a cosmetic and psychosocial problem and responsible for the popularity of schemes that claim to reduce body fat – but rarely do.

Plumpness with an android (apple-shaped) and, to a lesser extent, gynacoid (pear-shaped) distribution of fat plays a significant role in the causation of Type 2 diabetes and hypertensive heart disease. It also causes increasing insulin resistance, which brings its own problems. In these people weight loss is undoubtedly advantageous. Plumpness also modestly increases the risk of developing certain types of cancer but at the same time it protects against other causes of mortality. These compensate for increasing the risk of these life-threatening illnesses. As a result, the total mortality figures for plump people are similar to those for people who are lighter.

True obesity, in contrast to plumpness, should be looked upon as a genuine disease entity with many, largely unknown, causes.

It is impossible to define clinical obesity simply and exactly but a competent doctor should know it when he or she sees it. Obesity imposes extra strain on the heart, respiratory system and weight-bearing joints, and predisposes to snoring, poor sleep patterns, sleep apnoea and other morbidities. In the past, surgeons and anaesthetists avoided operating on obese patients for technical and clinical reasons. Modern keyhole procedures have made surgery – including bariatric surgery, which is surgery specifically designed to treat obesity – safe even in the fattest patients.

Prevention and treatment

Diet and drug treatments generally work for only short periods, if at all. Currently, the only reliable remedy available for the morbidly obese is a surgical bypass of the stomach, but, like all surgery, it is a poor substitute for a medical treatment that works.

While the availability of a plentiful supply of food is a prerequisite for the development of obesity, the relationship is far from simple. The normal spectrum of adiposity for human beings, defined as that associated with maximum longevity, becomes evident only when there is unrestricted access to food, such as now occurs in much of the developed world. It has a much wider range than was previously believed. People classified, in the past, as overweight to a degree that imperils their health are often found to have a normal life expectancy when more scientific criteria are employed.

The role of exercise in controlling body fat deposition has been recognised for a hundred years or so, but receives much less attention than dietetic control, mainly because it is less profitable. Exercise is important because of the way it affects bodily functions and not because of the calories it burns off. Obese people who carry around up to their own lean weight as fat need stronger and bigger muscles for the same degree of exercise as a slim person carries out. Consequently, if they lose weight simply by dieting rather than in combination with exercise, they lose muscle as well as fat. When, as almost inevitably happens, they then put on weight again, it is mainly as fat, accounting for the harmful effects of uncontrolled intermittent or recurrent dietary treatment of obesity. The importance of muscular exercise in controlling the propensity to become obese is exemplified by the high prevalence of obesity seen in patients with muscular dystrophy and other conditions that confine them to a wheelchair.

Moderate exercise, 20 minutes' brisk walk a day, is all that is required to keep the body healthy. It helps by keeping the tissue metabolism toned up.

Conclusions

The world's population has got fatter over the years as food supplies have become more readily available for more people. There is no reason to believe that this is detrimental for most people. With an increase in age of the population, an increase in average weight is not only to be expected but is probably beneficial. True obesity is a medical problem now, as it always has been. The difference is that in the past the very obese died young, whereas today they now survive, increasing the numbers alive at any time. They require medical treatment. The excessive and unnecessary attention given to the prevention of the misclassified condition of 'overweight' children and adults as 'obese' diverts attention from where it should be directed, namely the pathological condition of genuine obesity

The old adage that it is better to leave the dining table wanting more than to leave it fully satiated is probably still as good advice as any for those genuinely wanting to avoid gaining weight. This may be difficult to achieve in an 'I want it now' society, but it might be helped by teaching nutrition at an early age rather than resorting to propaganda based on half-truths and unproven ideas. But for many of the morbidly obese, the ones really at risk, it is only advances in the understanding of the pathology of obesity and its specific and appropriate treatment that offer any genuine hope of sustained benefit.

9

HEALTH FOODS & SUPER FOODS

VINCENT MARKS

PLAIN FOOD IS NOW rarely seen in the shops. Food has to be something more than ordinary: it has to be a *health* food, a *functional* food or a *superfood* – anything as long as it's not just ordinary unless, of course, its a 'wholefood'. Food can be additive-free, preservative-free, cholesterol-free, fat-free or sugar-free, with reduced or no salt or whatever else happens to be fashionable, such as trans fats. Food can be fresh, natural, organic or, if you are lucky, local, but never simply food. None of these many flowery descriptions tells you anything useful about whether it will taste nice, which is the main reason we choose to eat it. The other, even more important, reason is to relieve hunger and provide the energy and materials required to keep us alive and replace parts that are worn out. We need the proteins, fats, carbohydrates and minerals, as well as the vitamins and trace elements that enable us to use them, always in the right amounts and proportions.

How did these often meaningless descriptions of food get into our language, and do they serve any purpose? The short answer is that they do serve a purpose for the manufacturer, but seldom, if ever, for the consumer. The sellers get us to buy their food product rather than someone else's, but they do not really help consumers to improve their chances of living longer and healthy lives. The rest of this chapter will explain why.

Health foods

The notion that some foods are healthy implies that others must be less healthy or unhealthy. This may be true under certain circumstances. Even food that *is* healthy goes bad if it is not treated properly and so may become *un*healthy and become a bad food.

Food that doesn't make you ill is a healthy food – provided it is taken in a proper amount and as part of a balanced 'healthy diet'. Just because it is possible to overdose on a single healthy food doesn't make the food itself unhealthy. This applies to everything we eat. One of the most important properties of a food is the amount of it you eat – too much is just as bad as too little, but it takes longer to kill you.

Advances in food technology – including preparation, transportation and the increased use of both refrigeration and preservatives (to prevent bacterial and fungal contamination), together with a better understanding of hygiene and why foods become rancid – have all contributed to why the food we eat today is generally better than what we ate in the past.

The introduction of legislation controlling the production, quality and sale of foods was probably the largest single factor in improving the diets of the poorer members of the community.

Notwithstanding the statement 'Let food be thy medicine and

medicine be thy food' – attributed to Hippocrates – little thought was given to the role of food in the preservation of health before the middle of the 19th century. The first genuine evidence that certain foods might be especially healthful came from the observations and experiments of James Lind, a naval surgeon in the mid-18th century. His discovery was not acted upon immediately because it did not fit in with contemporary views of health and sickness. More than 40 years elapsed before Lind's discovery of the importance of fruit in the prevention of scurvy was recognised by the British Navy. It led to British sailors becoming known throughout the world as 'limeys', because they were issued, for economic reasons, with limes instead of the oranges and lemons that Lind had shown experimentally cured scurvy (he did not test limes). Fortunately, limes turned out to contain the same essential dietary element, ascorbic acid (vitamin C), as other citrus fruits, though at a lower concentration but at a cheaper price.

Lind's discovery was so poorly known or respected, even in Britain, that William Stark, a Scottish physician and pioneer of scientific (or experimental) medicine, died from self-induced scurvy at the age of 29 while conducting an experiment to prove that a 'pleasant and varied diet' was just as healthy as the simpler, stricter diets that many advocated at the time.

Physiology and pathology

Our present understanding of health and disease had to await the arrival of physiology and pathology, the sciences that underlie our understanding of good and bad health respectively. Various treatments had been proposed for scurvy but none was thought any better than another. Lind's adoption of the experimental method proved the benefit of one and the uselessness of the other six treatments that he tested experimentally.

Lind, a medically qualified ship's surgeon, selected 12 sailors suffering from equally severe forms of scurvy and divided them into six pairs. To each pair he gave one of the treatments that were in vogue at the time. Only the pair given fresh oranges and lemons responded by returning to good health within a few days of starting their treatment. Most of the others perished.

In the population as a whole, scurvy also disappeared. This was largely due to the growing popularity of turnips and, more especially, of potatoes as dietary staples during the 18th and 19th centuries rather than to citrus fruits, which were too expensive for all but the rich and middle classes until after World War II.

Perhaps the most important step in the understanding of the role played by food in health and disease was the demonstration that the classic view, that living things were imbued with vital forces that could not be described in ordinary chemical terms, was wrong. The synthesis of urea by Friedrich Wöhler in 1828 and the growth of what rapidly became known as organic chemistry laid the foundations for the scientific study and understanding of human, animal and plant nutrition. It should have put a stop to all the mysticism that surrounded food and drink, but it didn't, and the idea that the chemicals, of which plants and animals are made, differ from 'manmade' ones in some mysterious way persists. It is promoted by pseudo-nutritionists with no scientific background who attempt to differentiate chemicals of identical structure and composition into 'natural' and 'artificial', which by implication are 'unnatural', on the basis of their origin.

The ability and need to prepare, preserve and store food has always been with us, even when we were nomadic hunter-gatherers. Its importance increased as people started living in village communities and eventually large cities. The methods used, such as roasting, grilling, frying, boiling, milling, bottling, drying, salting and fermenting, to mention just a few, changed little from the beginnings of civilisation to the arrival of industrialisation of food production. This, together with improved means of distribution, occurred mainly during the 19th century, but in some cases, such as with sugar, tea and coffee, much earlier. They became the most valuable traded commodities in the world and fortunes were built on them. None of them was essential for good health but all of them improved the *quality* of life – hence their popularity. This was sufficient in itself to make them unpopular among those of a puritanical nature – an unpopularity that continues to the present day.

During the 19th, and to a lesser extent the 20th, century, the population of what are now called the developed countries increased enormously. In Britain, for example, the population more than tripled from 11 million to 32.5 million between the censuses in 1801 and 1901. Most of the growth was in the major cities; the countryside itself was becoming depopulated. In the typically rural county of Somerset, for example, the total population increased from 270,000 to only 430,000, most of which occurred in the towns.

The concentration of people in cities put an enormous burden on the supply of food, which would have been impossible to meet had it not been for improvements in the preservation, storage and transportation of food. Even so, illnesses that had previously been rare, such as rickets in children, and whose cause at the time was unknown but was later shown to be due to dietary deficiencies, became common, as did diseases such as tuberculosis, due to the rapid spread of infections in overcrowded conditions. Fresh food, such as milk, eggs, fruit and vegetables, was in short supply. The introduction in the mid-19th century of home deliveries of fresh milk, by the Express Dairy Company, was made possible by the rapid transport from the country to the towns by railway. This heralded a new era – but initially only for the comparatively wealthy. The city poor had then, as they still do, to rely largely upon factory-processed food. Refrigeration was the only way of preserving food in its native state.

The adulteration of food

With the development of food technology came the increasingly common practice of food adulteration. This had always existed but was difficult to detect. By 1820, however, analytical chemistry had developed sufficiently for Friedrich Accum, a German-born chemist living in London, to publish *A Treatise on Adulteration of Foods and Culinary Poisons*. In it he described some of the worst examples of food adulteration that had probably been going since people began living in towns and had to depend on merchants to provide them with food. The adulterants Accum dealt with ranged from such innocuous substances as water added to milk and beer, to the addition of chalk, alum and sawdust to flour, and the addition of such poisons as red lead to cheese to improve its colour. He not only described how the adulteration could be detected but also named those he had detected doing it. This caused sufficient reaction against him to force his return to Germany in 1821, where he lived a full and active life until he died aged 69 in 1838. It did not, however, bring an end to food adulteration.

Legislation had to wait another 55 years until the Sale of Food and Drugs Act – the first of its kind anywhere in the world – was introduced in 1875. It has been revised many times since then. Its efficacy depends upon enforcement by trading-standards officers using criteria established on the basis of advances in analytical techniques.

Opposition and objections

Advances in food technology in the preparation, processing and preservation of food aroused opposition from people who would nowadays be described as activists, entrepreneurs, religious fundamentalists or cranks depending on your individual viewpoint. One of the most notorious of these was Sylvester Graham, a Presbyterian minister whose message was that people should shun tea, coffee and alcohol, meat and the newly introduced white bread. They should instead eat wholemeal bread made from his particular brand of flour. He was either a crank or a shrewd businessman.

Another was John Harvey Kellogg, the inventor of cornflakes and founder of the company that bears his family name but that owes its success to his more realistic brother William, who didn't have the same religious hang-ups and was quite prepared to adopt advances in food technology.

Vegetarianism also made its Western debut in the mid-19th century having long been practised in the Far East by the Jains and other non-Christian religious groups. It owed its origins in the West, as in the East, to moral and religious objections to eating meat rather than to a belief in its health benefits. Although this is now one of its main selling points, the evidence that vegetarians are healthier and live longer than omnivores is suspect, to say the least, but, conversely, there is no reason to believe that it is harmful.

Not all of the opponents of the advances in food technology and hygiene that appeared during the 19th and 20th centuries were benign. The opposition to such an important development as the pasteurisation of milk, for example, was malign and would, had it been successful, have contributed to countless cases of tuberculosis and brucellosis, whose previously high incidence was dramatically reduced following its introduction. This simple manoeuvre had little or no effect upon milk's nutritive value and scarcely any on its taste, unlike sterilisation, which does. Despite being endorsed by such celebrities as George Bernard Shaw, the anti-pasteurisation brigade had little success when it began, though it still surfaces from time to time among those with little regard for evidence of health benefit and a lot for irrational belief in the safety of nature.

Whether people who wish to live recklessly and drink unpasteurised milk and cheeses made from it should be prohibited from doing so is a moot point. In the main, they are strong supporters of laws that are imposed on the food industry to protect the majority and with which they undoubtedly agree.

Essential ingredients

Lind's proof of the value of citrus fruits in the prevention and treatment of scurvy should have led to the conclusion that some foods contain essential ingredients that are absent from others. However, with the exception of salt, this realisation took years to occur.

The requirement for salt in a healthy diet, over and above its ability to enhance flavour and act as a preservative, had been known since Roman times and its issue to legionnaires is still commemorated in the term 'salary'. It was not, however, until the middle of the 19th century that the general concept of the necessity of including trace amounts of minerals in the diet was widely accepted and acted upon. It coincided with advances in agricultural chemistry and the need of plants for certain chemicals to enable them to grow properly.

Food grown on iodine-deficient soils in such diverse places as Derbyshire and Switzerland was shown to be associated with thyroid enlargements and cretinism, which could be prevented by the addition of iodine to table salt. This was a milestone in the development of nutritional science and preventive medicine and represents one of the first examples of food fortification. However, this did not reach its present proportions until after the discovery of what Funk, on a false supposition, called vitamins. Funk had identified a chemical carrying an amine group and subsequently known as thiamine (now generally called vitamin B_1) as the substance whose absence from the diet caused beriberi. This type of heart disease had become very common in the Far East and was linked to the substitution of polished rice for less highly milled rice. It had virtually disappeared by the beginning of the World War II, but returned with a vengeance among the Japanese prisoners of war. Funk correctly associated beriberi with other illnesses that were increasingly being linked to dietary deficiencies and forecast, incorrectly it transpired, that the chemicals lacking in them would also turn out to be amines.

The curious way that vitamins are named – and what dignifies some but not all substances given the name – owes much to the work of Gowland Hopkins of Cambridge. While a few substances present in food were erroneously believed to be essential, others, such as pangamic acid, were iniquitous marketing ploys.

The first 'health-food shop' opened in Britain in 1894. It bore little resemblance to the majority of 'health-food shops' of today, which are stuffed full of 'natural' and synthetic vitamins, herbs, mineral and diverse specious food supplements, many of little or no nutritional value. What it would have contained were foods that could be bought in ordinary grocery stores. Its unique selling point was that it only sold foods that could be described as 'natural' and free from adulteration by dyes and other manmade additives of dubious quality and safety. This, remember, was at a time when food adulteration was still rampant and long before the rigorous health-and-safety testing of today. It would also have stocked the new foods, now universally called breakfast cereals, that were in fashion in America, and gradually replaced the traditional fried breakfast of the English and porridge breakfast of the Scots.

The problem with health-food shops then, as now, is that there is no evidence that so-called 'health foods' are any better than foods bought in conventional food shops; or that, just because they are 'natural', they are safer or more nutritious. To describe a food as 'healthy' implies that other foods are unhealthy. This use of the term 'healthy' has not remained the prerogative of health-food shops or the health-food industry as a whole. Regrettably, it has been adopted by all manner of food producers and marketeers and increasingly by governments. To reinforce the concept of some foods as healthier, because they had undergone less processing than others, terms such as 'wholefood', 'organic' and 'natural' have been employed. Initially introduced by people who presumably genuinely believed that these various properties confer some nutritional advantage to the consumer, they have cynically been increasingly used by one branch of an organisation that sells 'ordinary food' in another.

Freshness

Food (and drink) contaminated with bacteria or moulds, or food that has become rancid, undergone adverse or harmful chemical changes as a result of processing, ageing or storing is likely to be deleterious to health and is now forbidden by law. The freshness of food is probably one of its most important and healthful properties. It is, however, difficult and in some cases impossible to achieve except for those living in small self-contained communities or those who keep an allotment. The invention of refrigeration and then of onsite freezing brought fresh fish, vegetables, fruit and meat to the masses but, like other advances in food technology, it has not been without its critics, who, despite evidence to the contrary, claim that chilled and frozen food is less nutritious than food consumed under ideal conditions. Even if it were true, the impracticability of conveying food from its origins to its consumers in a pristine state must be taken into account.

The opposition of vocal but scientifically uninformed pressure groups has kept food preserved by irradiation off the British market, although it is freely available in other equally health-conscious countries. Fears that food preserved by irradiation becomes radioactive and consequently dangerous, instead of beneficially free from bacterial or fungal infestation, are rife. This fallacy is encouraged by pressure groups who either should know better or do not trust the regulatory authorities to distinguish between food that has gone bad but rendered bacteriologically safe by irradiation from food that is preserved when fresh and kept in that state for shipment to distant markets.

The Sale of Food and Drugs Act 1875 was the beginning of the inspectorate's ever-growing concern for quality as well as quantity.

The Act empowered inspectors to test food and drugs for adulteration. From it developed the complex legislation that now controls the composition, purity and labelling of foodstuffs. In the main, mankind has learned, by trial and error, what plants and bits of animals can or cannot be eaten safely. Nevertheless, toadstools, mistakenly picked as mushrooms, are still an important cause of illness and death in some parts of the developed world. Similarly, people who eat rhubarb leaves or green potatoes, when food is scarce, become ill or die. Because this sort of historical evidence is not available for new foods, most developed counties, such as the UK, have introduced legislation covering what are called *novel foods*, which are introduced either from abroad or by invention. These have to undergo more thorough safety testing than traditional foods, some of which are not always as safe as they are presumed to be.

Today the consumer can be confident that any food purchased from an authorised source is 'healthy'.

10

THE FOOD WE EAT

VINCENT MARKS

DOGMA
Natural and organic foods are good for you because
they do not contain chemicals.

THERE IS A WIDESPREAD belief that farm-grown produce
and organic foods are better for you than others because they do
not contain chemicals used in large-scale conventional farming. It
is this that has given rise to the idea of 'good foods' and 'bad foods'.
There is a belief that compost and other so-called organic fertilisers
are good because they are not made up of chemicals that are
harmful or that there is a difference between the chemicals they
contain and those applied directly to the soil.

   This dogma is wrong. All plant nutriment comes from the air, in
the form of CO_2, and from water-soluble chemicals in the soil. The
composition of these chemicals is the same whether they come from
a plastic bag or from manure or compost. It is the magic of nature
that turns these simple chemicals into the food that we eat. Without

them the food chain would fail. It is these chemicals, whether originally from a natural or a processed source, that provide the material for the carbohydrates, fats and proteins of our diet.

To understand why this is so requires a knowledge of the nature and composition of the food we eat and what happens to it when it is processed by our bodies.

In spite of all the hoo-hah that surrounds it, there is nothing mysterious about the food we eat; it is just another complex collection of chemicals. So is drink, which is just a special kind of food. What makes food unique is that we eat or drink it to the exclusion of everything else except medicines. The idea that any food has medicinal properties and provides a special immunity from disease is fanciful and has no basis in fact.

In ancient times, and right up to the early 19th century, food was thought to contain vital forces that could not be manufactured. The new era of organic chemistry, including the chemistry of food, began with the synthesis of urea in 1828. It proved that a natural chemical could be made artificially out of laboratory chemicals. Organic chemistry (unlike inorganic chemistry, which is the study of minerals and metals) is concerned with the nature of substances based on carbon – just another form of the carbon in carbon dioxide, with which climatologists and politicians are currently preoccupied.

The growth of analytical organic chemistry in the 19th century laid the foundations for the scientific study and understanding of human, animal and plant foods and nutrition. It should have put a stop to the mysticism that had previously surrounded food, but it didn't – at least not completely. The idea has been nurtured among certain groups that the chemicals that make up plants and animals differ in some mysterious way from 'manmade' ones. As we saw in Chapter 8, such ideas are promoted by pseudo-nutritionists, with no scientific background.

By the middle of the 19th century, the three main energy-supplying constituents of food – the proteins, fats and carbohydrates – had been identified, as had most of its mineral components.

Carbohydrates are so called because they are composed exclusively of carbon, hydrogen and oxygen in the same proportions as those found in sugar. Fats, too, consist only of carbon, hydrogen and oxygen but in a different proportion from those in carbohydrates and, on a weight-for-weight basis, yield twice as much heat when metabolised in the body. Proteins, on the other hand, contain not only carbon, hydrogen and oxygen but also nitrogen and sulphur.

All three classes of energy-supplying nutrients are composed of relatively simple building blocks. Carbohydrates are made up of monosaccharides, fats of fatty acids and proteins of amino acids. Irrespective of its source, all the food we eat must be broken down into these components in the gut before they can enter the bloodstream.

Carbohydrates

Carbohydrates are the most plentiful organic compounds on Earth. They are made up of simple sugars. By far and away the most important of these sugars is glucose. Starches and cellulose are both made up exclusively of glucose molecules linked together to form chains of varying length and complexity. Both starches and cellulose can be broken down into glucose. Mammals use enzymes to break down starches in the gut but they cannot digest cellulose.

Humans lack the enzymes necessary to break down cellulose but they are found in the bacteria that live in the intestines of herbivorous animals. That is why herbivorous animals, such as the cow and the elephant, can live on foods consisting mainly of cellulose, whereas we cannot. Because human beings cannot digest cellulose, it constitutes roughage or dietary fibre, or, to give its modern name, non-starch carbohydrate.

Glucose, derived from starch in bread, potatoes and rice, is our principal source of carbohydrate, although there are many other sources in our food. The next commonest carbohydrate is fructose, which, like glucose, is a simple sugar. Though both are composed of six carbon atoms, twelve hydrogen atoms and six oxygen atoms, the exact way they are all joined to one another is slightly different. Because fructose, in solution, bends a beam of light to the left, it is described as *laevorotary*, whereas glucose bends it to the right and is described as *dextrorotary*. This difference led to fructose being called *laevulose* (left-bending) in contrast to glucose or *dextrose* (right-bending).

Fructose, is most plentiful in plant foods. It occurs, joined to glucose in sucrose, which is table sugar. Before sucrose can cross from the intestine into the blood it must be split into its components: glucose and fructose. The two simple sugars are then absorbed by different mechanisms involving the cells lining the intestine.

The only other important dietary simple sugar is *galactose*. It is joined to glucose to form lactose. The sole dietary source of lactose is in milk. Some babies are unable to split lactose into glucose and galactose and, as a result, the lactose remains in the intestine, where bacteria ferment it. This produces gas, which causes bowel distension and pain. It is a cause of infant colic. In the most severe, but rare cases, it can lead to severe disability or death.

Since virtually all carbohydrates enter the bloodstream as glucose or fructose, it is nonsense to consider one form of the parent carbohydrate from which they come to be good and another bad. Once in the bloodstream the metabolic effects of glucose and its effect on bodyweight are the same irrespective of its source.

Fats

Fats are mostly composed of fatty acids linked to glycerol, a chemical relative of carbohydrates, to produce *triglycerides*. Triglycerides cannot be absorbed into the body until the fatty acids have been split off from the triglyceride by enzymes in the intestine. Once inside the cells lining the intestine, the fatty acids are recombined to form triglycerides and linked to a carrier protein. These combined molecules, the so-called *lipoproteins*, carry the fat into the lymphatic vessels and around the body and, finally, into the bloodstream.

Fatty acids are chains of carbon atoms with one end of the chain linked to oxygen. The length of the chain varies from 2 carbon atoms (acetic acid) to 26 or even more. The nomenclature of the fatty acids describes how many carbon atoms they contain and how they are joined to one another. The carbon atom next to the one at the end that is linked to oxygen (the acid part) is called *omega*, the one next to it is *carbon 2 omega* and so on.

Some fatty acids are *unsaturated*. This means that some of the carbon atoms are linked to their neighbour by a double bond. This makes them more chemically active than those linked by a single bond. Those without double bonds are known as *saturated fatty acids*.

The number of double bonds in a fatty acid varies enormously. Those that have only one double bond are called *monounsaturated* and those that have many are *polyunsaturated*.

The position in the chain where double bonds occur is important from a nutritional point of view. When a double bond exists between carbon atoms 3 and 4, it is described as an *omega-3 fatty acid*; those between carbon 6 and 7 as *omega-6 fatty acids*. Although human beings can make most fatty acids from carbohydrates and fats, they cannot produce double bonds in these positions. Because fatty acids containing them are essential constituents of all our cell membranes and in the prostaglandins (chemical messengers important in inflammation), omega-3 and omega-6 fatty acids are described as *essential fatty acids*.

Information on the nature of fats has accumulated slowly and the fact that some fatty acids are essential whereas most are not only came to be recognised during the late 1980s. This accounts for the present interest in them. They are, however, required only in trace amounts, just like vitamins, and are not ordinarily a major source of calories.

The pioneers of food chemistry distinguished between fats and waxes, both of which share the property of not mixing with water. Waxes occur more commonly in plants than in animals, but cholesterol, the one that attracts the most attention, is confined to animals and never occurs in foods of plant origin. Plants may, however, contain chemicals very similar to cholesterol that can interfere with the absorption of cholesterol and, under certain circumstances, actually cause a fall in plasma cholesterol concentration.

Proteins

Proteins are made up from a large number of different amino acids joined together to form a chain, which can then rearrange itself in a variety of ways to give each protein its own specific characteristics. These are altered by cooking as well as by other physical processes. Although there are theoretically tens of thousands of amino acids, virtually all proteins in the body are made up from just 20 of them. The order in which they are arranged and the shape of the chain they form distinguishes one protein from another.

Amino acids are made primarily by plants from five elements, namely nitrogen and sulphur, which come from chemicals taken in through the root systems, and oxygen, hydrogen and carbon. Animals cannot make amino acids – they have to get them through their food. They do, however, have the capacity to change some amino acids into others so that, of the 20 or so amino acids commonly found in proteins, only about 11 are essential. The amount of these essential amino acids present in food determines its nutritive value, especially during growth.

Food proteins that contain all of the amino acids in the proportions that they occur in the human body are complete, or first-class; those that do not are incomplete, or second-class. The proteins in milk, meat and other foods derived from animal sources are generally complete, whereas those from plant sources seldom are. This disadvantage of a vegetarian diet can easily be overcome by ensuring that the diet contains a variety of different proteins from many different plants. It is perfectly possible to have a good and complete diet made up of foods of only vegetable origin – but it is more difficult.

Whatever the original source of the protein, it must first be broken down in the gut into its constituent amino acids. Even so, their absorption requires the active participation of the cells lining the intestine to transport the amino acids, from the lumen of the gut into the blood. In some, very rare hereditary diseases, the machinery for transporting individual amino acids is lacking. This causes illnesses linked to specific amino-acid deficiencies. Ordinarily, however, dietary protein deficiency is a consequence of too little first-class, or complete, protein in the diet.

Proteins are essential for the body since we cannot manufacture the amino acids they contain. Amino acids are the building blocks of the tissues of our body. Proteins from animal sources contain more essential amino acids than those from plants. No matter what animal meat is consumed, it is a better source of the proteins we need than plants.

Minerals

Minerals, mainly metals – calcium in bones, for example – were recognised as an essential component of the body long before the origins of organic chemistry. In a massive tome on the chemistry of the body and diet and its role in the prevention and treatment of disease, Jonathan Pereira, an academic English physician, listed the elements that had been found to be present in the body and that, he thought, must also have been present in the food.

A great number of minerals are found in the body. The function of some, such as iron, calcium, sodium, potassium and iodine, are well known; others, present in very small amounts, such as selenium and cobalt, play a vital role, but are necessary in only minute quantities in order to maintain long-term health.

Water

The human body is made up largely of water – some 60 per cent by weight in men and 55 per cent in women. It is less in women because, weight for weight, they contain more fat, which contains the least water of all.

We need about 2 litres (3.5 British pints) of water a day in order to produce 1–1.5 litres (1.8–2.6 pints) of urine, and it doesn't matter whether we drink it or take it in as food. The water content of ordinary food varies from about 15 per cent in butter to 96 per cent in celery or cucumber. Under temperate conditions, the average person gets about 20 per cent of his or her daily water requirements from food, even if it doesn't contain soup, which can be up to 98 per cent water.

So poorly understood is the chemistry of food by the majority of people that, as recently as 2006, a self-proclaimed expert was reported as saying that soup doesn't contain water. Like many of his ilk, he suggests that only fluids labelled as water are able to meet our daily needs. This misunderstanding is widespread and accounts, at least in part, for the current practice of using a bottle of water as a fashion accessory and the notion that water in solid food and drinks such as tea and coffee 'doesn't count' towards our daily requirements

Vitamins

The sixth group of essential dietary constituents were discovered as recently as 1912. Their existence was neither known nor even suspected, except by the most advanced thinkers of the time. Sadly, the implication of the observations of James Lind (whom we met in Chapter 9) on scurvy, in 1750, was not appreciated at the time. This made all the attempts to produce artificial diets doomed to failure. It is only within the past 30 years or so that knowledge of all our nutritional requirements, including vitamins, has been sufficiently well established to enable people to stay alive for 20 years or more on wholly synthetic diets, delivered intravenously.

There is no evidence that anyone in the developed world, eating a normal mixed diet, is likely to suffer from a vitamin deficiency or likely to benefit from additional supplementary vitamins. The exceptions are those not exposed to sunlight who might lack vitamin D or those on a vegan diet that lacks Vitamin B_{12}. Taking an excess of some vitamins may be dangerous. Any excess of vitamin C, a common additive, is immediately excreted in the urine.

11

LABELLING & LEGISLATION

VINCENT MARKS

Substances added to food

SUBSTANCES HAVE BEEN added to foods to make them more attractive, more palatable and above all more profitable. In the past many have had little, if any, nutritional value. In more recent times substances have been added to food to improve their nutritive value – a process generally known as *fortification*. The Food and Drugs Acts need constant updating as the technology available to enforce them improves and knowledge of nutrition, pharmacology and toxicology determines their value.

If the substances added have no nutritive value, they can be considered adulterants; some may be harmful. Adulteration of food, including drinks, is as old as antiquity. As we saw in Chapter 9, it was brought home to modern society most forcibly by the publication in 1820 of Friedrich Accum's book *A Treatise on Adulteration of Foods and Culinary Poisons*.

Consumer legislation in the UK had to wait until the Sale of

Food and Drugs Act of 1875. This was the first of its kind anywhere in the world but was soon copied by every other developed country, including the USA, which introduced its Food and Drugs Act in 1906. Among those most influential in getting legislation passed were the English physician Doctor Arthur Hassall and the American novelist Upton Sinclair, whose book *The Jungle* exposed the disgraceful practices rife in the meat industry in Chicago.

Toxins already present in food

Poisons have been known since Biblical times and were scientifically investigated long before medicines. Most of the potent poisons are derived from plants; they are 'natural' products.

It was not appreciated until well into the 20th century that even the most wholesome of foods might contain toxins that were harmless in small amounts, but with continued usage might predispose to cancer or other chronic diseases. Some of the classical associations that have been made between food and disease in the past, such as that of port with gout and of potatoes with spina bifida, are known to be wrong, while others, such as persistent alcohol drinking during pregnancy and consequent brain damage in the children, are less well known but true.

It is important to remember the words of Paracelsus, written 500 years ago: 'All food and drink is poison, it is a matter of the dose.' With few exceptions, such as true allergy, this fact indicates the importance of considering the amount of food eaten when discussing the potential to do harm.

Food labelling

The perceived role of food in causing disease has changed dramatically following the involvement of governments. Often government intervention is the result of representations from powerful pressure groups and many of the resulting restrictions are opinion-based, rather than factual.

When the nature of food was obvious from its appearance, or could be explained in just a few words by the vendor, labelling was unnecessary. The increasing industrialisation of food production and the complexity of its composition, together with changes in retailing practice, have made labelling necessary. It is controlled by legislation, which has gradually evolved from the early Food and Drugs Acts, which were primarily concerned with the safety and quality of foods and medicines sold to the public.

Very few foods are eaten without having undergone some process to alter the chemical nature of one or more of their constituents to make them more palatable, more digestible and less toxic. Cooking alters the chemical nature of the constituents of many stable foods. Potatoes and red beans, for example, become edible only after cooking. Raw potato starch is indigestible until cooked. Cooking red beans destroys the lectins they contain, which are poisonous in their native state. Cooking not only alters pre-existing constituents but it can also generate new ones – some of which are toxic, such as acrylamide, while others are flavoursome and safe, such as the chemicals that produce the aroma of roasted coffee.

No system of food labelling is completely satisfactory – nor, given the complexity, is it ever likely to be. Early food labels attached to prepared foods sold in cans, bottles and packets gave information on the nature and relative proportion of the ingredients and in many cases how much protein, fat and carbohydrate they contained. Gradually, additional information was added and regulated by food-labelling legislation. Some ingredients, especially additives, have to be specified by name or their E number which is a seal of approval from the EU and indicates it has been passed as non-toxic in the doses used. Other additives could be described as permitted flavours.

At the end of the day, it is the totality of what has been consumed rather than the amount of each actual ingredient that determines whether it is beneficial or potentially harmful.

Eating food is more than just the intake of a mixture of essential dietary requirements – it would be extremely dull if it were. Even among the most deprived members of society, the offering and sharing of food with others is a basic social activity and, for the individual, eating and drinking should be a pleasurable event.

Food today contains a variety of substances that give it colour, flavour and texture, all of which add to the pleasure of eating. Many also contain pharmacologically active ingredients, such as caffeine in tea and coffee, and substances that are poisonous but are natural constituents of plant foods. It is because we take in only a limited amount of these foods that we can safely ignore their potential harmful effects.

The chemical nature of food and why it is nutritious appears to be poorly understood by the public. In recent years, iconic nutrition labelling has been introduced in an effort to inform the public, but is a very poor substitute for proper education. Iconic labelling is far from perfect. Much of it is based on such mantras as 'sugar, salt and (saturated) fat are bad for you', despite the lack of convincing evidence. Labelling can do as much harm as good when incorrectly applied or when the implication of the label is misinterpreted or poorly understood. For example, anchovies contain high concentrations of fat, salt and occasionally traces of minerals such as cadmium; however, it is unlikely that anyone would consume enough anchovies to affect their health. Nevertheless, they will be labelled as if they were a danger.

Many food writers and food gurus use public ignorance as to the true nature of food to promote a particular belief or fashion. This is then taken up by food manufacturers and retailers to promote a new product or to claim some exclusive virtue of an old one. They use emotive slogans, such as 'natural', 'wholefood', 'organic', 'pure' and 'non-GMO'. These labels should be treated with caution. Often they are nothing more than a marketing ploy to give a product a temporary advantage, which, if it catches on, is soon adopted by others.

The use of labelling to allow consumers to choose foods that yield fewer calories, in the belief that by eschewing those foods with a high calorie content they might lose weight, ignores the main problem of obesity. It is not so much what one eats but the *amount* of it that causes one to put on weight.

Understanding food legislation

The discipline of food science achieved academic status only during the mid-20th century. It studies all aspects of food and drink production, composition, storage and transportation, Together with the science of nutrition, it provides the objective basis for regulatory legislation.

The advances made in these two sciences since the days of

Accum, Periera and Pasteur are enormous, but much remains to be done, including eradication of the prejudices expressed by those who erroneously believe that food was better or healthier in the 'good old days of natural food'. We now know that the ingredients of foods, whether prepared at home or mass-produced, undergo chemical changes during their preparation, storage and transportation. A worldwide panic occurred when it was revealed, in the 1990s, that many of our most popular prepared foods contain acrylamide, a known carcinogen. This serendipitous finding resulted from improvements in analytical technology that allowed the detection of one part per 10 million. Its application to home-cooked foods revealed a similar presence of acrylamide in them, too. Fortunately, it occurs at so low a concentration as not to constitute a risk to health.

Some foods are well known to contain pharmacologically active ingredients, some of which, such as phyto-oestrogens, can have profound effects on the body. A few foods contain substances known to be toxins or poisons when taken in more than minimal amounts, and they serve no dietary purpose. Oxalic acid is one such substance, which is present in many of our most delicious fruits and foods; it is also produced in our own bodies, but it is poisonous if taken in excess. Cyanide is found in trace amounts in almonds, and the high level of potassium found in bananas could prove fatal to patients with renal disease.

The supply and distribution of food

For most of man's history food has been in short supply. In the past far more people went to bed hungry than satiated; about a quarter of the world's population still do, in spite of the enormous increases in food production that have come about as a result of the application of science to agriculture and fisheries. The reasons for world hunger go far beyond the inability to produce enough food and drink to meet everyone's needs. Due to political and financial considerations, there is mal-distribution of produce, especially in Africa. Waste and the spoilage of food due to poor storage, vermin and contamination are estimated to account for up to 30 per cent of the food produced in the developing world. In developed countries, an almost equal amount is wasted and ends up in the bin.

Conclusion

It is only by a better knowledge of the food we eat and the physiology of nutrition that we can understand why those who claim that the food produced by one particular method of farming is better for us than another, or that a particular food will prevent ageing and disease, are wrong. There is no scientific basis for their belief.

A ban on the use of chemicals and pesticides in food production, which some advocate, would reduce the worldwide availability of food. This would be most keenly felt by those in the less developed world. It would make all of us vulnerable to the effects of fungal contamination, such as ergot on cereals and aflatoxin on peanuts; organic farming would reduce the agricultural yield by up to 30–40 per cent. If widely adopted, it would be a disaster in a world experiencing difficulty in feeding an expanding population.

A ban on GM foods would lessen the amount of land that could be farmed and reduce the yield. It may make the crops more likely to be attacked by parasites and moulds. It would result in attempts to reduce blindness by preventing the farming of vitamin-A-rich golden rice and other valuable foods.

Those who promote organic food and spread the fear of conventionally produced food and of GM crops must accept the potential damage that their unsubstantiated, unscientific dogma may cause.

12

WHAT REALLY CAUSES HEART ATTACKS?

MICHAEL SMITH

> For every complicated problem, there is a solution
> that is simple, direct, understandable and wrong.
> —H. L. Mencken

DOGMA
High cholesterol causes heart disease.

Almost every day, on opening a newspaper, we see a 'scare story' about society or our personal wellbeing. Many of these stories are repeats of longstanding threats that gain greater currency by being repeated and not challenged. GM (genetic modification) technology became 'Frankenstein foods', which was a brilliant but misleading headline and became embedded in the public mind. Certain additives that fell into a similar category were highlighted as likely to be dangerous, poisonous or somehow morally wrong. It was left to the countries where there wasn't the luxury of navel gazing and fine words to grow GM crops and to produce ample amounts of food, often with extra benefits, such as saving the sight of countless numbers of children by giving them vital vitamin A to prevent corneal damage.

The big stories, often driven by pressure groups, committed academics and politicians, are currently the 'epidemic of obesity', global warming, individual carbon footprints, all of which may have some factual basis but there is a lurking feeling that the simple arguments and the pejorative language used exaggerate the dangers and may possibly be wrong.

A lesson from history

Medicine is not immune to 'big ideas'. Years ago it was obvious that gastric and duodenal ulcers were caused by too much acid 'burning' holes in the gut lining. If only the output of hydrochloric gastric acid could be reduced or even eliminated, the problem could be contained and the ulcer would be allowed to heal. Gastrointestinal research departments everywhere strived to show that the maximum acid output (stimulated by an injection of histamine) in an ulcer patient was greater than in non-ulcer control subjects. Using sophisticated statistics and subgroup selection, clinicians tried hard – sometimes too hard – to explain the persistence of ulcers, especially in stressed individuals. Antacids and acid-blocking drugs could keep the symptoms at bay, but, when these failed, surgeons designed more and more elaborate operations to eliminate the acid threat. But the root cause remained a mystery until, in Australia, a trainee pathologist published a letter in the *Lancet* commenting that the base of all the ulcers he studied after surgery showed the presence of rod-shaped bacteria called *Helicobacter pylori*. It was shown, by physicians in Australia, that a few days' treatment with a cocktail of antibiotics and an acid-blocking drug could eliminate an ulcer completely.

It took about ten years for the idea to enter fully into practice in the UK, but now it is almost the only treatment apart from the use of modern antacid therapy in people who have not responded. Modern trainee surgeons seldom see or perform gastric surgery except for bleeding or perforations of ulcers commonly associated with the use of steroids, nonsteroidal anti-inflammatory or aspirin-based drugs. A further benefit of elimination therapy was the reduction in gastric cancers. A single voice from the inquisitive open mind swept away decades of research, surgery, diets and medicines.

The cholesterol/heart attack story

For over a decade, many cholesterol sceptics have been pointing out the flaws in the currently accepted views on why in some of us our arteries 'fur up' with fatty deposits that can block the flow of blood, nutrients and oxygen to heart muscle, but not in others.

The cholesterol story began after World War II, when it became apparent that the incidence of heart attacks in the USA was rising remorselessly and causing deaths in relatively young men.

A physiologist and nutritional expert, Ancel Keys – the man responsible for producing the so-called K rations for the armed forces and later for the starving populations in post-war Europe – did a study of fat imports and usage in a number of countries. He compared the number of deaths from heart attacks with the level of fat consumption. His seven-country graph was startling. Finland ate the most fat and had the greatest number of cardiac deaths; Japan had the lowest fat intake and had the smallest number of deaths; the USA was in the middle. A similar relationship was claimed for the amount of the cholesterol in the blood and the incidence of heart disease. Later, when the whole study of all the countries involved was published, the apparent strong correlation between high fat intake, cholesterol in the blood and deaths from heart disease disappeared. However, the idea of a dietary causation of cardiovascular disease (CVD) had taken hold.

The World Health Organisation supported a study of 27 countries with 113 academic monitoring sites ('Monitoring of Trends and Determinants in Cardiovascular Disease' or 'MONICA') and the findings were published in 1989. The results were very similar to the Keys findings. The highest mortality for heart disease (Finland) was five times greater than the lowest (France) at the same cholesterol level. In Karelia, north Finland, the mean cholesterol was 6.3mmol/dl with deaths from coronary heart disease (CHD) at 493 per 100,000 inhabitants. In France, for a similar cholesterol level, the mortality from CHD was only 102 per 100,000. Overall, the study failed to show any relationship between cholesterol and CVD except for Japan and China, who both showed low heart disease and low cholesterol. In Japan, it was a disgrace to die of a heart attack and it was more acceptable to have the death recorded as a stroke. This may well have distorted the observations.

Examination under the microscope of furred-up arteries in the heart and elsewhere in the body containing blemishes to their ordinarily smooth lining showed the presence of increased amounts of cholesterol. So it became 'gospel' that a high-fat diet and hence more cholesterol in the blood was the culprit.

The message went out that special diets were necessary to reduce these noxious substances. Once, while dining in a research laboratory café in a Los Angeles hospital, an American colleague looked at his fried egg and carefully dissected the yolk away from the white albumen. He looked up and said, 'Goddamnit, Mike, I don't know which bit I shouldn't eat.'

He was a total convert but a confused one for this diet/heart hypothesis. The whole edifice linking cholesterol in the diet to heart disease should have collapsed in the USA when a several-hundred-million-dollar, seven-year study known by the initials MRFIT (multiple-risk-factor intervention) involving 360,000 middle-aged men from 18 cities was started by the American National Heart, Lung and Blood Institute in the USA. They selected 12,000 men who seemed to have a higher risk of CVD than the others and divided them into two groups. Six thousand had intensive checks and treatment, for example blood-pressure treatment and detailed dietary advice on a low-fat and low-cholesterol diet. Serum cholesterol was measured before and after the trial and morbidity and mortality figures in both groups were collected from their physicians. The second group, however, did not have any special management but were asked to return annually for checks. After seven years, the results were analysed and proved to be disappointing for the researchers, who had expected that the diet/heart hypothesis would be proved and that the gospel could be confirmed and preached.

Risk factors such as smoking and high blood pressure were found to be associated with an increased risk of heart attacks but it was so in both groups. There was only a tiny, 2 per cent difference in cholesterol levels (which is not statistically significant) and the number of deaths from CVD and total mortality (all causes) were virtually equal in both groups. A subgroup analysis showed that the only difference in death rates was between smokers and non-smokers. As the dangers of smoking were already known, it was an expensive, but useful, negative trial.

That should have been the end of the fat and cholesterol story. Many other trials have repeated the work but none has shown a clear correlation between dietary fat and CHD.

The most famous field study was started in 1948 in a Massachusetts town called Framingham. The population was regarded as sufficiently stable for long-term observation. After 16 years of close monitoring, 193 individuals had died of heart attacks and 1,378 of the population remained free of CVD. There was very little difference in cholesterol findings between those who died and those who survived. A later analysis after 30 years for men between 48 and 57 years showed a similar poor correlation. The men were divided into four cohorts depending on their starting cholesterol (Group 1: less than 4.6mmol; Group 2: 4.6–5.6mmol; Group 3: 5.7–6.7mmol; Group 4: more than 6.7mmol). At all times, during the 30-year follow-up, there were more deaths from heart disease in the 4.6–5.6mmol group than in the higher-cholesterol 5.7–6.7 cohort.

The cholesterol theory of heart disease fails to explain why, since the 1960s, the incidence of heart disease has fallen steadily in New Zealand, Australia, the USA and Canada with no change in fat consumption, which has remained steady at 40 per cent.

The finding that many patients with CHD do not have a raised cholesterol level has been observed many times. It suggests that the 'epidemic' of heart disease is not due to too much fat and cholesterol in the diet.

The Department of Health certainly still believes the fat-in-the-diet idea, as do the media and the supermarkets, with their low-fat everything. It is only gradually being accepted that it is wrong by doctors and the public

The pharmaceutical industry, which makes blood-cholesterol-lowering medicines, including statins, has a vested interest in keeping the cholesterol story going, as it is a far easier one to hammer home than explaining what statins actually do and the dangers they present.

In *Panic Nation* (edited Feldman and Marks), Dr Malcolm McKendrick, in a chapter entitled 'The Great Cholesterol Myth', debunked the cholesterol/heart story and likened it to a magnificent cathedral built on shaky ground.

Two recent books on the subject, one by U Ravnskov, *The Cholesterol Myths – exposing the fallacy that saturated fat and cholesterol cause heart disease*, and *The Great Cholesterol Con* by Dr McKendrick, deal very persuasively with all the research and scientific evasions that pervade the cholesterol/heart disease story.

What *really* causes coronary heart disease?
What really happens when arteries age and become patchily thickened?

The heart is supplied by three large arteries. On one side of the heart the left ventricle, which pumps blood at pressures rising to 140mm Hg in middle life, has two arteries, which arise from a single vessel coming from the aorta. The low-pressure right ventricle pumps blood only to the lungs and works at about 16mm Hg. The arteries branch and branch again to cover all the cardiac muscle.

WHAT REALLY CAUSES HEART ATTACKS?

A heart attack occurs when one of the branching arteries becomes blocked. This can be due to reversible spasm, which is more common in women and smokers, and may give pain due to lack of oxygen, but can reverse in time to allow the muscle to recover. If, however, the artery is suddenly blocked by a clot forming on top of a patch (plaque) of atheroma, then the muscle downstream will be killed off producing what is known as a *myocardial infarct* – the commonest cause of a heart attack. The administration of a modern clot-dissolving drug given in a hospital A&E department or in an ambulance, within 30 minutes of the pain starting (fibrinolysis treatment), may prevent the heart muscle being damaged.

Not all clots cause complete obstruction to the flow of blood and it is now thought that a plaque of atheroma may rupture many times allowing an insidious closure of the channel. In this case the heart has time to open up collateral or bypass channels, which can take over the blood supply to the threatened area. If the high-pressure left ventricle is threatened, all that someone would know is that they have bouts of angina, a tight feeling across the chest and down the left arm and possibly some shortness of breath. If damage occurs to the low-pressure, right ventricle, it may go almost unnoticed.

What is striking is that atheroma mainly occurs at branching points of the arteries. Until a few years ago, atherosclerosis was thought as a slowly progressive, degenerative disease predominantly affecting older persons. Now it is realised that it is a more dynamic inflammatory lesion and its early stages can be seen in young people and is usually reversible. The cellular and biochemical mechanisms are becoming increasingly understood and are beginning to swing the management pendulum from mechanical surgical interventions by stents (tubes inserted into blood vessels) and bypass operations to medical treatments.

Atheroma begins with damage to the inner lining of the artery called the *endothelium*. The endothelium produces nitric oxide (NO). This 'gas' has many protective roles, which include preventing the formation of free radicals (single oxygen atoms that are highly destructive to cellular DNA) and preventing clotting factors accumulating in the lumen of the vessel. The first sign of trouble is a reduction in NO. This can be due to smoking, diabetes, severe alcohol intake, high blood pressure or chronic infection (the bacterium chlamydia has been found in atheroma). It occurs most readily at points of mechanical stress such as where the artery branches. There may also be a genetic predisposition to endothelial failure.

It is possible that high levels of fats and cholesterol may diminish NO production and set the scene for inflammatory and lipid invasion of the arterial wall.

When nitric oxide is reduced, the cells lining the artery secrete 'sticky' molecules that attract and capture inflammatory cells in the blood and move them into the wall of the artery, which produces an atheromatous plaque. It is currently believed, however, that it is oxidised low-density lipoprotein (LDL) that encourages the onward progress of the inflammation – hence the postulated advantage of antioxidants about which we hear so much.

As the inflammatory cells in the embryonic plaque take up fats from the blood, they enlarge and change into foam cells. They are then filled with cholesterol, much of which is made *in situ*. This is the beginning of the atheromatous process. The danger comes from the addition of inflammatory proteins, which may cause the plaque to rupture its cap.

Some plaque ruptures produce only a fine tear and clot formation is limited. Anticoagulant drugs such as aspirin and warfarin can limit the extent of any fissure. Highly complex antioxidants are believed to participate and should theoretically reduce the amount of oxidised LDL that perpetuates atheroma formation but, so far, dietary trials with antioxidant-rich food have been disappointing.

To summarise, according to Professor Peter L Weissberg of Cambridge, England:

- atherosclerosis is an inflammatory reaction to the subendothelial accumulation of modified fat;
- atherosclerosis is invariably associated with abnormal endothelial cell function; vascular smooth muscle cells are the only cells capable of protecting against rupture of the 'cap' and its thrombotic consequences;
- the outcome of atherosclerosis is determined much more by plaque composition than by plaque size;
- atherosclerotic lesions frequently enlarge as a consequence of repeated subclinical episodes of rupture and repair; the only logical conclusion to be drawn from the angiographic (x-ray studies of the blood vessels) and outcome studies of statin treatment is that the drug stabilises plaques;
- atherosclerosis is a dynamic, inflammatory process capable of being modified.

The statins

The group of drugs with the generic name *statins* are interesting and certainly do more good than simply 'lowering your cholesterol'. They have become the 'wonder drugs' of cardiovascular medicine, but is the hype justified? It is fair to say that reducing blood cholesterol is the least of the effects of statins and it would be fairer to the public perception of them if their real benefits were highlighted. It could at least save yet another trip to the GP or hospital to have a cholesterol check and yet another exhortation to eat less fat or low-cholesterol food.

The first statin, mevastatin, was isolated from the fungus *Penicillium citrium* by two Japanese researchers, Akira Endo and Masao Kuroda, and for which Akira Endo was awarded a Lasher Prize in 2008. The American company Merck & Co. followed up their work and produced lovastatin in 1976 from another fungus, *Aspergillus terreus*. And so the commercial pharmacological story began.

Statins work at the very beginning of the biochemical pathway that leads to cholesterol synthesis in liver cells, and also in other specialised cells. It is this part of the body's synthesis of cholesterol that is blocked by statins. As a by-product of statin activity, the liver cells produce more LDL receptors, which remove LDL from the blood, and so further reduce the blood cholesterol level.

Statins reduce cholesterol synthesis. They also reduce the ability to make the inflammatory proteins and the clotting factors that are the immediate cause of a heart attack.

One of the statins, pravastatin, reduces cholesterol by 60 per cent and this may be why plaque size is reduced with statin treatment – but it takes up to five years to achieve meaningful changes.

However, not all statins show uniform results. Before 2002,

when the British Heart Foundation reported in the *Lancet* there had been six major trials of either lovastatin or pravastatin lasting five years or more. Three of the trials showed very little effect on cardiovascular outcomes, i.e. fatal heart attack or non-fatal disease. Two other trials showed a reduction in the risk of non-fatal cardiac attacks and lower incidence of strokes. The biggest impact came from the 4S (Scandinavian Simvastatin Survival Study). It showed for the first time a lowering of the risk of both fatal and non-fatal CHD. The cry went out from the press conference held to launch it to 'lower patients' cholesterol now'.

The improved survival figures for the patients on simvastatin must be weighed against the very large numbers who took the drug without any benefit.

Statins do have a place in cardiovascular medicine but the benefits are delayed for about one year. Thereafter, there is a slight but increasing reduction in heart disease and stroke. After five to seven years, the size of the plaque shrinks slightly but the lumen volume remains unchanged. The value of these drugs has been excitedly promoted but there are some reservations about their efficacy. They have 'feet of clay' and are not the universal panacea for heart disease.

13

METALS & MYTHS

ANDREW TAYLOR

DOGMA
We are being poisoned by metals in our food
and in the environment.

DID HENRY VIII HAVE SYPHILIS? This is a question that has long intrigued historians. One aspect of the debate reflects on the likely treatment that syphilitic patients of the 16th century would have received. It can be argued that, if the king had been diagnosed with this disease, his physicians would have given him mercury, just like other sufferers. Mercury is a toxic metal and some of the effects of exposure to it include the increased production of saliva. This was used by doctors to judge when they were giving their patients the right dose. If Henry did have syphilis, he would have shown excessive salivation, which would surely have been obvious to members of the royal court. Contemporary records, particularly the reports sent by foreign ambassadors, fail to reveal any evidence for this, suggesting that Henry's famous moods and behaviour were probably not syphilis, as has so often been stated in the past.

The harmful nature of metals was very well understood long before Henry's reign. It was known to Pliny, the Roman historian, who records that miners extracting metal-rich ores used face masks, constructed from thin animal bladders, to prevent inhalation of potentially lethal dusts. The inadequate provision of some metals is just as important and potentially harmful as an overexposure to them at work or from medicines containing them.

In bygone days, livestock in certain areas were noted to be so unhealthy that they did not produce good-quality meat. Considerable efforts were made during the first half of the 20th century to determine the causes of these problems. Eventually, it was recognised that poor-quality meat came from animals that were grazing in regions where the soils, and consequently the plants growing on them, had low concentrations of certain essential elements or minerals. When animal foods were supplemented with minerals such as the copper and selenium that were lacking, the meat quality and farmers' profits improved. Not until many decades later was it recognised that signs and symptoms similar to those exhibited by these animals could also be found in humans. These observations led to our understanding the importance of essential trace elements, such as metals, in health and the fact that they can also be poisonous if taken in excess.

Metals and health

The term 'trace elements' was originally applied to elements present within the body at concentrations too low to be reliably measured or even detected in healthy individuals. With the development during the 20th century of much more sensitive instrumentation, it became possible to determine the presence of virtually all elements at infinitesimally low concentrations. A useful working definition of a trace element is one found at a concentration of less than 0.01 per cent of the body's dry weight.

Some trace elements are recognised as having vital physiological and biochemical functions in the body and are described as being *essential*. If they are present in inadequate amounts, bodily functions start to go wrong and produce signs and symptoms of disease and even death. Conversely, when the body is exposed to an element beyond its capacity to utilise it, this produces a different set of diseases and may lead to death or chronic invalidism. These essential elements and also those not known to be useful in any way to the body will become toxic if exposure is increased beyond the safety limit. In fact, some of the essential elements, such as selenium, are more toxic on a weight-for-weight basis than other notoriously 'toxic elements' such as arsenic. Some elements capable of causing harm are regularly prescribed as pharmaceutical agents.

When it comes to the myths of diseases and calamities associated with metals, it's usually the result of excessive exposure rather than a consequence of a deficiency, that attracts attention. As with King Henry, it is mercury that currently features most prominently in these stories. It is important to realise that mercury comes in several forms, each with its own chemical and biological properties. Most of us are familiar with metallic mercury, but there are also inorganic mercury compounds, such as mercuric chloride, and organic compounds, such as methyl mercury. Inorganic compounds of mercury are rarely found except in specialised industries and it is the mercury metal and the organo-mercury compounds that are of particular interest to clinicians and toxicologists.

Mercury

Mercury, as the liquid metal, is quite innocuous and can be swallowed without harm, though it can be absorbed through the skin when rubbed into it (as it was when used as a treatment for syphilis). With just a small increase in ambient temperature, liquid mercury produces an invisible vapour, which can be inhaled and absorbed through the lungs to produce signs and symptoms of disease. In addition to the increased production of saliva, the effects of exposure to inhaled mercury include inflammation of the gums, a fine tremor affecting fingers and other parts of the peripheral nervous system and changes in mood and personality characterised by features such as intense blushing, tearfulness and temper outbursts. There are also reports of modestly increased blood pressure. In contrast to these symptoms, poisoning by organic methyl mercury primarily involves the central nervous system, where the main effects are disturbances in vision, speech and balance. In pregnant women, methyl mercury affects the foetus, causing malformations and foetal death.

There have been many accounts of poisoning following gross exposure to both mercury vapour and methyl mercury. Does this mean that at much lower doses they will produce more subtle but still real effects? With very little evidence to go on, there are those who have promoted the idea that unexplained neurological disorders, such as motor-neurone disease, child-developmental delay and autism, are caused by mercury released from dental amalgam or by the regular consumption of fish and other types of seafood, the main source of organic mercury. Can this be true, or is it just another one of the myths surrounding diet and health?

Methyl mercury in seafood

Superficially, the arguments to support the claims relating to eating fish and mercury toxicity appear quite convincing. Methyl mercury – a by-product of certain manufacturing industries – certainly accumulates in predatory fish such as tuna. We have measured an increase in blood mercury concentrations in people eating such fish on a regular basis. There is a child in the UK who insisted on eating swordfish for almost every meal and showed symptoms of poor coordination, which resolved when he was persuaded to alter his diet.

However, while such examples are most unusual and their cause unproven, there are communities where seafood is regularly eaten in large amounts. Two substantial studies have sought to determine the influence of mercury in seafood eaten during pregnancy – and in the children's own diet as they became older – on the early growth and development of infants. In one of these long-term studies, in the Faroe Islands, whale meat was the source of the methyl mercury, whereas various fish species were consumed in another project on subjects who lived in the Seychelles. In both groups, the intake of methyl mercury was higher than typically seen in most other populations but there was no detrimental impact on the growth or neurological development.

Mercury in dental amalgam

What about the mercury in dental fillings? Mercury in amalgam does vaporise, especially when the surface is rubbed or comes into contact with hot drinks. There is also a relationship between the number of amalgam surfaces and the concentration of mercury in urine. What is unclear is the actual amount normally released and consequently available for absorption into the bloodstream.

Ingenious experiments using devices such as a mechanical 'mouth' have been developed to answer this question. With this equipment, it was found that a single amalgam filling provides 0.03 micrograms/day, which is almost 3,000 times less than the safety level permitted in the USA for persons with occupational exposure to mercury and is far too small to be responsible for any symptoms. Those who remain convinced of mercury's involvement with these disorders are not dissuaded by this information. It is claimed that the metal slowly accumulates in the tissues, where it has a deleterious effect, and that this accumulation is not reflected by concentrations in blood or urine. Demonstrating the validity of these arguments by its proponents is achieved by a variety of dubious tests, such as chelation challenge and the analysis of hair, neither of which withstands critical scrutiny.

Chelation-challenge tests

Chelating agents are used to treat people who have metal poisoning. These drugs take hold of metal atoms from inside tissues and draw them back into the blood so that they can be eliminated in the urine. This property has been exploited to devise tests to diagnose situations where metals have accumulated because of a genetic disorder, such as Wilson's disease, a condition where copper accumulates in the body, especially within the liver and brain. If Wilson's disease is suspected, the patient is given a dose of a chelating agent called penicillamine and the amount of copper excreted in the urine in the next 24 hours is measured.

From this well-established and scientifically proven test, other regimes have been developed. They have been applied by practitioners of dubious probity to subjects who complain of feeling unwell and whose symptoms are attributed to metals in their tissues. In these 'tests', urine is collected before and after administration of a chelating agent with the ability to bind to almost any metal. They are unlike penicillamine, which is more or less specific for copper. The inevitable increase in urinary excretion of various metals, such as mercury and lead, that follows this procedure is then claimed to confirm a fictitious diagnosis of metal poisoning. Following the 'test', the patient is advised to embark upon an expensive course of treatment. The remedies recommended are generally supplied by the organisation that carried out the test in the first place and include further courses with the chelating agent to remove the 'accumulated' metal from the tissues.

There are two distinct flaws to these procedures. Unsurprisingly, since we all have some metals that are derived from our food, water and general environment in our tissues, anybody taking the test will show a positive response. Furthermore, some of these laboratories express their results in a way that will tend to exaggerate the amount present in the 'test' sample. A woman who had experienced difficulties in becoming pregnant came to see me. After eventually conceiving, she had been persuaded to have the chelation test, which apparently showed high levels of mercury in her tissues. She was extremely anxious and was seriously considering having a termination until reassured that she had nothing to worry about. I recently saw another young woman who said she had spent more than £30,000 on chelation tests and treatment for mercury toxicity, even though she never had any amalgam fillings.

Studies are designed to evaluate scientifically the effectiveness of chelating tests. In one, the response in a group of individuals who said that they were unwell because of mercury from their fillings were compared with a similar group without any symptoms. The increased concentration of mercury in the urine was exactly the same in both groups.

In a second study, the normal increase in excretion in individuals who were healthy was compared with the increase reported by a testing laboratory. The normal increase was no different from the so-called 'diagnostic' response. This study also demonstrated a problem with this type of testing. Some of the volunteers had a reaction to the chelating agent and became quite unwell. This effect caused the investigators to abort the study before all the subjects had been tested. Unfortunately, this was not an isolated case. A child with autism, another condition said by some fringe practitioners to be caused by metals in the tissues, died from receiving chelation treatment.

Hair analysis

A GP contacted me for advice as to what could be done to help one of his patients who had sent a sample of her hair to a testing laboratory and among the 'abnormal' results was an increase in the amount of chromium. The interpretation provided by the laboratory included the note that chromium causes cancer. The patient was worried. What should he do?

This is a typical example of pseudoscientific quackery. Hair analysis can be of value when there has been a suspected extreme exposure to trace metals, as in a murder inquiry, but it is of no value in assessing deficiencies of essential elements or to detect small increases in exposure or uptake. The concentrations of several elements are influenced by the colour of the hair, the age, gender and race of the individual so that simple reference or 'normal' ranges have limited relevance. There are some elements – aluminium, for example – where no increase is found even when tissue concentrations are grossly elevated. Conversely, normal or even increased values can be obtained when there is a dietary deficiency.

Hair will also trap elements from the external environment on to the external surface and washing procedures can be either ineffective or too aggressive and cause loss of metal from within the hair itself. Falsely high results are seen after shampoos. An interesting example of external contamination comes from the analysis of hair taken from Napoleon when he was exiled in St Helena. Medical reports describe him as having episodes of acute abdominal pain prior to his death, which could be consistent with arsenic poisoning. When hair samples were analysed, they revealed high concentrations of arsenic, leading to speculation that he was murdered. However, an autopsy was performed the day after he died and a gastric tumour was found, indicating that he died from cancer. Further investigations showed that the wallpaper of the room in which he spent much of the time before he died contained an arsenic-based pigment, which is quite volatile and is likely to have permeated the air, causing his hair to have the high levels of arsenic that were found many years later.

Ethnic remedies and food supplements

Lead in the environment caused concern because of its known toxicity. Lead additives in petrol were held to be responsible not only for impaired brain development in children but also for disorders such as multiple sclerosis, behavioural problems, delinquency and criminal behaviour. Following a series of elegant studies, it was shown that there is a small effect on IQ, amounting to one or two points, associated with environmental exposure to lead. To put this into a sensible context, the effect is similar to that on IQ that can be attributed to birth order of the child among his or her siblings. As a consequence of these investigations, leaded petrol was phased out in many countries. In the UK, a large group of healthy individuals volunteered to have their blood lead concentrations measured annually for two years before and for two years after the removal of lead from petrol. The results showed a steady fall in blood lead levels, even though atmospheric levels were unchanged. The change in blood lead levels probably had more to do with improvements in food quality and hygiene than to a direct effect of changes in airborne lead.

Emissions of lead from motor vehicles undoubtedly contributed to the amount of lead in food but many other factors were clearly involved in the decline of blood lead that have continued to be evident long after the complete abolition of leaded petrol. Nevertheless, cases of lead poisoning are still seen from time to time. The usual situations we encounter are where old, leaded paint is being sanded when families are redecorating their homes, where cosmetics or ethnic remedies have been imported and when individuals have the habit known as pica.

Most laboratories with an interest in trace elements have a collection of medicines that they have analysed and found to contain huge amounts of lead (and often other metals). These 'medicines' are usually from the Indian subcontinent, Middle Eastern countries and, more recently, from China. We have found dangerous amounts of toxic metals in remedies being sold for the treatment of diabetes, impotence, intestinal problems, lack of energy and teething in children.

Organic foods

With a trend developing for increased consumption of 'organically produced' foods, it is relevant to consider whether these are necessarily any more 'healthy' than conventional foods as far as metals are concerned. In one study on this subject, the concentration of the toxic metal cadmium was significantly higher in the kidneys of 'organically reared' pigs than those from pigs that were conventionally fed. When the diets were analysed, there was no difference in the amounts of cadmium present and it was concluded that the source was probably the 'organic' vitamin/mineral supplements they were given. In a second investigation, less iodine – an essential element – was found in 'organic' milk compared with conventional milk.

We have carried out our own investigations and looked at ten elements, seven of which are essential, in a variety of meats, fruits, vegetables and dairy products. For most food types, the concentrations of the essential elements and those with known toxicity were similar in 'organically' and 'conventionally' produced samples. However, the mercury content of 'organic' beef and cream was lower than in their 'conventional' counterparts. Similarly, cadmium content was higher in some conventionally produced vegetables. On the other hand, we found that nickel, an essential element, was present at lower concentrations in the organic foods. Conventional beef had higher concentrations of all the ten elements than organic meat, an observation that may be related to the supplements given to these animals. While these differences were real, they were not very large and almost certainly not clinically significant. We concluded that organically produced foods offer no nutritional advantage over conventional foods, as far as providing a higher intake of essential trace elements or lower intake of harmful non-essential ones, is concerned.

Along with organic food, dietary supplements are widely available and popular. Most people eating a sensible diet have no need to increase their intake of essential elements. In excess, they can cause an impairment in the intestinal absorption of other essential metals. Exceptional use of zinc supplements, for example, inhibits the uptake of copper and we have seen cases of profound copper deficiency produced in this way. There is one situation, however, in which a dietary supplement may be valuable. Selenium taken on a regular basis, at an amount a little greater than the recommended dietary allowance, was shown, in at least one well-conducted clinical trial, to delay the onset of, or prevent, prostate cancer.

Degenerative disease and metals

Trace elements have been incriminated in the development of some degenerative diseases. This has focused on two substances: one a metal, aluminium, the other silicon, a non-metal whose essential dietary status is still undetermined.

If aluminium reaches the brain, it is extremely neurotoxic. Sadly, there was a time when patients on kidney dialysis accumulated aluminium from the dialysis fluids used to treat them. They developed a condition that was named *dialysis dementia*, from which many died. Increased amounts of aluminium were found in their brains at postmortem. Some of the signs and symptoms from which they suffered were similar to those seen in Alzheimer's disease, which led to speculation that aluminium might be involved in the aetiology of this disorder. As a consequence there have been recommendations to avoid aluminium kitchenware in order to preserve brain function. Is this sound advice?

Experiments to measure the release of aluminium from containers into food were first carried out a century ago in 'The Lancet Kitchen'. Results from these early studies have been confirmed by more recent work using much more sensitive analytical tests. Foods that are acidic or alkaline will dissolve aluminium. The majority of foods, however, show little or no increase in aluminium content after cooking or on storage in aluminium containers.

Of course, there is still the question about acidic rhubarb and apples that have been cooked in an aluminium saucepan. At this point, we can also ask about people who habitually use aluminium-containing antacid tablets. Do they develop dementia? The answer is no. There is no evidence to suggest any association between dietary intake of aluminium and Alzheimer's. Aluminium is very poorly absorbed from the intestine – less than 0.1 per cent of the aluminium ingested. What is absorbed into the blood is promptly and efficiently excreted by the kidney (except, of course, in those without any – the reason aluminium proved toxic to those on dialysis).

Like aluminium, silicon is one of the most abundant elements in our environment and, as with aluminium, the gut is extraordinarily effective at preventing all but a tiny fraction from being absorbed. However, there are occasions when this barrier is bypassed and the amount of silicon in the body is increased. This occurs in women with a silicone breast implant. The traces of the silicone polymer leaking out of the implant have been implicated as a cause of arthritic joint dysfunction. Several careful studies were undertaken in which silicon was measured in the blood and urine of sufferers and a control group, and then compared. Another study compared the incidence of symptoms in patients with implants with a control group. In neither type of study was there any evidence to support this particular myth.

Summary

Some metals are essential to life; others have valuable pharmacological properties and are used to treat disease; but all can produce harmful effects on excessive exposure. It is right, therefore, to be cautious when considering how metals in our environment interact with physiological and biochemical processes that sustain life. We have immensely powerful tools that allow us to measure both exposure to metals and the effects that may follow. We can and should make use of these tools to give us the evidence upon which to base judgements as to how and when metals, and other elements, are responsible for beneficial or harmful effects.

14

THE NEXT GREAT PLAGUE?

RODNEY CARTWRIGHT

Influenza has been known in Europe, as an epidemic disease, since the middle of the 13th century. Notices are preserved of six visitations in the 14th century, and of four in the 15th; but these accounts being mainly those of the ordinary chroniclers of the period, are necessarily very meagre and imperfect.

SO BEGINS THE BOOK entitled *Influenza or Epidemic Catarrhal Fever of 1847–8* by Thomas Peacock MD, published in 1848. In 1920, the Chief Medical Officer presented a report to the Minister of Health in London on the 'Pandemic of Influenza, 1918–19'. In his introduction he states,

This document deals with one of the great historic scourges of our time, a pestilence which affected the well-being of millions of men and women and destroyed more human lives in a few months than did the European war in five years, carrying off upwards of 150,000 persons in England and Wales alone.

There is not a person who is unaware of influenza and the fear that the world may see another outbreak that will, in spite of modern medicine, cause untold deaths and suffering. Influenza is a viral disease that can cause small outbreaks confined to a community, larger outbreaks or epidemics involving a wider area such as a country, and global outbreaks, or pandemics, that spread across the world in a few weeks or months. Written records describe outbreaks of respiratory illness comparable with influenza infection stretching back over the centuries. No one knows exactly when the next pandemic will occur, but there are few epidemiologists who do not believe that time is drawing near when the influenza virus will again cause many deaths across the globe. The situation has been concisely summarised by Edgar Marcuse from the University of Washington School of Medicine with the words: 'The pandemic clock is ticking – we just don't know what the time is.'

The World Health Organisation (WHO) forecasts that the next flu pandemic will lead to the deaths of 2–7 million people worldwide with the number of persons affected running into billions. It is estimated that in the United States alone some 90 million will become sick and 2 million will die. The numbers are smaller in the UK due to our population size, but the Department of Health is nevertheless planning for at least 15 million to become ill and 56,000 to die. The time span for all this to happen is measured in weeks or a few months. A 'tabletop' exercise in the UK postulates that, when the pandemic starts, the first case will occur in late December, with 3 million cases by mid-March. It is reckoned that 3,000 people will die.

Is this all scaremongering whipped up by the media? Is there any evidence that this will happen, and, if so, surely modern medical techniques will be effective, won't they?

What is influenza?

There are few people who do not believe that they have had 'the flu' and some are convinced that they have it often, and especially after a flu vaccine jab. But what is flu? To most people it is a feverish illness with aches and a chesty cough. It is true that these symptoms may be due to the virus influenza but they can also be caused by a myriad other viruses that may cause an 'influenza-like illness'.

True influenza is due to the influenza virus, which comes in two main types: influenza A and influenza B. Both produce infections that range from being asymptomatic through to something much worse than the common cold. Its main distinguishing features are a feverish cough of rapid onset accompanied by a general malaise, muscle aching, headache, sore throat, runny nose and, not infrequently, some diarrhoea. Those mainly affected are children and the elderly. Outbreaks occur every winter to a greater or lesser extent but the numbers are limited. Outbreaks due to influenza B tend to be milder and geographically limited to smaller areas such as a town or district.

Influenza A is a complex and variable virus causing both limited outbreaks and worldwide pandemics that in the past have been responsible for severe illnesses with deaths. A relatively benign strain of the influenza A virus can develop major changes in its structure, enabling it to acquire the properties necessary for it to become a pandemic strain such as the ones the world witnessed in 1848 and 1918–19.

Thomas Peacock, in 1848, describes a disease in which:

The attack was most usually sudden, the patient experiencing a sense of cold down the back and between the shoulders, lapsing into general chilliness or complete rigors, and succeeded by flushes of heat and dryness of the skin, pain in the head, chest, and extremities, and prostration of strength.

The report of the 1918–19 pandemic states,

In most cases the patient had been ill for a day or two with ordinary simple influenza, not necessarily severer than that of his neighbours, when there was a rapid or sudden change for the worse, and the picture changed rapidly from that of influenza to that of a chest case; and the effects of the pulmonary changes were often so fulminating that death might ensue in 24, 36 or 48 hours.

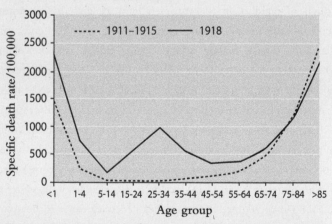

Figure 14.1: Pneumonia and influenza mortality by age in the United States as recorded in the US Vital Statistics Rates for 1900–40. The dotted line is the average for the years 1911–15. Specific death rate is per 100,000 of the population in each age group.

The severity of the infections was not the only important observation in the 1918–19 outbreak. Another aspect of this pandemic that distinguished it from normal outbreaks of influenza was the age profile of those who died from the infection. In the years preceding the pandemic most of the deaths occurred in those under 2 years or over 65 years with few deaths in young adults. During the pandemic, there was a striking increase in deaths in the 15–35 age group, an observation that has been made in succeeding influenza pandemics.

So not only do the pandemic strains spread widely, causing many deaths, but they also have the ability to be fatal in otherwise healthy young adults. I personally witnessed this during the 1968 flu pandemic. A healthy teenager undertook his paper round in the morning, returned home then went to school. At lunchtime he felt unwell and was feverish. He was sent home and his mother put him to bed. Later in the afternoon he had a very high temperature, had breathing difficulties and was blue or cyanosed. He was admitted to hospital but died shortly afterwards. The autopsy revealed changes in the lungs that were so well described in the 1918–19 report:

> The lung lesions, complex or variable, struck one as being quite different in character to anything one had met frequently in the thousand of autopsies one has performed during the past 20 years.

The lungs showed, in varying degrees, acute congestion, diffuse haemorrhage, bronchopneumonia, haemorrhagic infarcts and collapse.

The reasons why infections due to pandemic strains can result in such severe, overwhelming disease, especially in healthy adults, is only just being understood. One possible reason is that the pandemic strains are potent inducers, by the infected tissues, of chemicals called *pro-inflammatory cytokines and chemokines*, which produce a cytokine storm. This affects the lungs adversely. The cells lining the respiratory tract respond vigorously with infiltration by inflammatory cells, tissue destruction and flooding with fluid that exudes into the spaces normally filled with air. The patient begins to drown in his or her own fluids. Unless this is recognised and treated immediately, it results in the lethal outcome seen in more severe flu viral infections, especially those that cause fatalities among young healthy adults.

The proposed mechanism of the cytokine storm evoked by the influenza virus can be viewed in an animated form on http://www.cytokinestorm.com/cytokine_storm.html.

The complexities of influenza A

Influenza A is a complex virus that has the ability to alter its structure, enabling it to overcome any immunity resulting from exposure to previous strains. When someone is infected by influenza A, the body responds by producing antibodies that neutralise the virus and at the same time prime the body so that, if the same virus enters the body again, it can be rapidly neutralised and no infection result. This is the natural immunity that develops following an infection. Immunity can also be achieved by the administration of vaccines, but they are effective against only the specific strains represented in the vaccine.

In humans, clinical influenza has mainly been due to strains of the virus containing H1, H2 or H3 haemagglutinins and N1 or N2 neuraminidases in varying combinations. As time progresses, both the haemagglutinin and neuraminidase structures of the flu virus can alter so that the body's defences, primed by previous infecting strains, no longer recognise it. The changes may be minor and the structure is said to have 'drifted', or they may be major, resulting in more significant outbreaks of infection. The new haemagglutinin or neuraminidase is given a new number and, to all intents and purposes, appears as a new virus against which the population has no immunity. This means that no protection exists and a major outbreak or pandemic can occur.

The 1918–19 pandemic was due to an H1N1 virus, that of the 1957 pandemic to H2N2 and the 1968 pandemic to H3N2. The H1N1 strain reappeared in 1977, but by this time the number of people who had experienced the 1918–19 pandemic was small, so there was little population immunity to it.

Influenza A also infects many animals and birds, providing a 'breeding ground' for new strains. The 1918 strain has been traced back to pigs in the USA. The recent interest in influenza and the concern that a new pandemic may be near has largely been due to the emergence of a new strain in birds, H5N1, which I will discuss below. Some animal strains may cause infection in humans but the severity of such infections and their ability to spread between humans is low.

It is thought that, if a human or animal becomes infected with two strains of influenza A at the same time, some swapping or reorganisation of genetic material may occur so that the emerging virus is a combination of the two infecting strains. This may be a mechanism leading to a shift in a virus's virulence. Alternatively, a mutation in the haemagglutinin or neuraminidase may occur spontaneously, resulting in a new type.

What are the prerequisites for pandemic influenza?

For a pandemic to occur, it is necessary for a number of factors to exist at the same time. A new influenza subtype must arise with a haemagglutinin unrelated to strains that are already circulating in the population; there must be little or no pre-existing immunity; the new subtype must cause clinically apparent disease and must spread efficiently from person to person.

Is there any sign that these factors are developing or coming together? In 1996, a 'new' strain of influenza, H5N1, was recognised in a farmed goose in China, followed by an outbreak in Hong Kong the next year. The spread of H5N1 within birds, both domesticated and wild, has been detected in Asia, Africa and Europe (see Figure 14.2). The World Organisation for Animal Health had recorded cases in birds in 22 countries including the United Kingdom by April 2008.

This avian strain appears in two forms: one that spreads rapidly through a flock causing a high level of mortality in a few days, and a second type that spreads slowly and is less lethal. Already it is known that animals can become infected with fatal results, as was seen in zoo tigers fed on chicken meat that may have been infected.

In 1996, 18 human cases were reported in Hong Kong with a fatality rate of 33 per cent. There then appeared to have been a quiet period, although this may be due to lack of reports rather than the lack of cases. The WHO has monitored the developing situation closely. As of 18 March 2008, a total of 373 human cases had been confirmed with 236 dying, a 63 per cent mortality rate. Cases have been reported in 14 countries of which the majority are in Asia (see Figure 14.3).

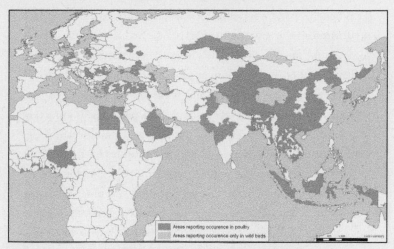

Figure 14.2: The global distribution of reported cases of H5N1 influenza in poultry and birds since 2003 (Source: WHO)

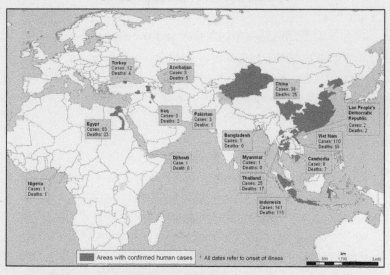

Figure 14.3: The global distribution of reported human cases of H5N1 influenza since 2003 (Source: WHO)

When the world was poised for a pandemic of bird flu, news came from Mexico of a new strain of H1N1 derived from a pig strain – swine flu hit the headlines. It has been characterised by human-to-human spread of the virus into at least two countries in one WHO region. This is WHO Phase 5 and while most countries are not affected at this stage, it is a strong signal that a pandemic is imminent and that the time to finalise the organisation, communication and implementation of the planned mitigation measures is short.

This new strain has not yet been associated with a high mortality rate but it is recognised that the same pattern was observed in the early stages of the 1968 pandemic. The second wave in the autumn was associated with severe disease. Will the same apply in 2009? Will the H1N1 swine flu be the next 'great plague'? Only time will tell but the world is prepared in a way not previously seen.

Modern medicines and influenza

Surely, with all the advances in medicine, are there not effective ways of combating the new pandemic strain? The two main weapons in the medical armoury against viral infection are vaccines and antiviral drugs.

Vaccines

Influenza vaccines historically have been targeted against the haemagglutinins and neuraminidases of the current circulating strains of influenza. New strains require the development of a new matching vaccine, a process that takes a minimum of four to six months. The production of the vaccine in large quantities then takes a further unquantifiable period of time. The question then arises as to who should receive the vaccine in the short term and

176

who in the longer term. Global immunisation is not feasible, so choices will need to be made and nations will need to determine their priority populations. At present, we do not know the exact makeup of the next pandemic strain, although the considered opinion is that it will be either an H1N1 or an H5N1 variant. Vaccines have been prepared against the avian strain but there is no certainty that they will be effective against an emergent human pandemic strain.

The UK Department of Health has stockpiled 3.3 million doses of the yet unlicensed pre-pandemic H5N1 vaccine. This is to be used only if the risk of a pandemic arising from H5N1 increases significantly and then, initially, only for healthcare workers and essential service staff. It is possible that a new type of vaccine, currently undergoing clinical trials and targeted at a protein called M2e, will offer some hope of reducing the impact of a new pandemic strain of influenza. M2e has not altered over many pandemics and vaccines against it may prove more useful than those directed against haemagglutinin or neuraminidase.

Urgent action is being taken to develop a vaccine from the H1N1 swine strain.

Antiviral drugs

Since the last pandemic of influenza, a number of antiviral drugs effective against influenza A have been developed. They fall into two main groups: the *neuraminidase inhibitors* and *amantadine derivatives*.

The neuraminidase inhibitors, zanamivir and oseltamivir, block the active site of the influenza viral enzyme neuraminidase and so reduce the number of viruses released from an infected cell. They have been shown to reduce the duration of illness from influenza and also to reduce its severity, provided treatment is started within two days of the onset of symptoms. The existing stockpiles of antivirals are sufficient to allow for the treatment of all symptomatic patients and arrangements to make it rapidly available are a critical part of the health response. Of concern, however, are reports of resistance to the drugs in some patients infected with the avian H5N1 strain. A big question mark therefore remains as to how effective these drugs will be in a pandemic situation.

The amantadine derivatives (amantadine and rimatadine) are thought to interfere with the influenza A virus M2 protein by inhibiting virus replication and resulting in decreased viral shedding from infected cells. These drugs have been used with good effect in seasonal outbreaks of influenza A. Tests on the effectiveness of the amantadine derivatives against various strains of influenza have, however, shown that resistance to them is developing, so that they are no longer recommended as first-line therapeutic agents against influenza.

All this means that, although recent advances in vaccines and antiviral agents may help to prevent and control the severity of influenza in individuals, there is no magic bullet that will have a major beneficial effect on the population as a whole.

This is well recognised by governments and institutions throughout the world, resulting in the development of detailed pandemic contingency plans not only to manage those who become ill or die but also to maintain the structure of society as we know it.

Never has any disease that has not yet arrived resulted in so much forward planning involving all parts of society. Perusal of the contents pages of the national framework for responding to an influenza pandemic published by the Cabinet Office and the

Department of Health in January 2008 provides sobering reading.

After setting the scene by describing seasonal and pandemic influenza, it makes various planning assumptions. Depending on the virulence of the influenza virus, the susceptibility of the population and the effectiveness of countermeasures, up to half the population could have developed illness and there could be anywhere between 50,000 and 750,000 additional deaths. All this would be expected to take place over a period of only a few weeks. In order to maintain the functioning of society, many different government departments will be involved and all aspects of life will be affected. It will indeed become a national and international emergency.

If, as would be expected, a large number of the population become ill over a short period of time, there will be dramatic effects on the functioning of the country's infrastructure. All parts of society will be affected, including those responsible for essential services such as electricity generation, the supply of water and food, the maintenance of law and order and so on.

The economic effects will be considerable with illness-related absenteeism from work estimated at a cost of £3 billion to £7 billion, excess mortality costing a further £1 billion to £7 billion and the loss of future lifetime earnings at £21 billion to £172 billion. Globally, it has been estimated that the cost of a pandemic to the world economy will be in excess of $800 billion or 2 per cent of the global GDP over a whole year. The Cabinet Office has encouraged all businesses to develop their own pandemic plan to ensure the continuation of their business. Advice on factors to be considered are to be found on the UK Resilience website (http://www.ukresilience.info/pandemicflu.aspx).

Similar advice has been given by the US government to its

commercial and business sector. The International Monetary Fund has prepared a paper entitled 'The global economic and financial impact of an avian flu pandemic and the role of the IMF' in order to assist the financial sector in their contingency planning.

Conclusions

A pandemic of influenza – will it or will it not be the next great plague? If it will, when will it explode upon the world? Will the world be ready? Will it disrupt society as we know it, even if for only a few months? Will all the preparations and plans against a virus strain that as yet does not exist provide the means for influencing the expected consequences of a pandemic? Will the measures proposed do more than exacerbate the differences between the highly technically advanced countries and those that still struggle with the basic requisites of life?

In spite of all the advances of modern medicine across the globe, we can but wait, watch, plan and hope that the observations expressed by Sir Winston Churchill in 1890 in his poem 'The Influenza' are not to be repeated. Perhaps the stanza of the poem that best paints a picture of an influenza pandemic is the one that reads,

> O'er miles of bleak Siberia's plains
> Where Russian exiles toil in chains
> It moved with noiseless tread;
> And as it slowly glided by
> There followed it across the sky
> The spirits of the dead.

15

ELECTRICITY AND LEUKAEMIA

DAVID JEFFERS

DOGMA
Power lines cause cancer.

Reactions to pylons in the environment

ELECTRICITY IS VITAL to modern society but few welcome its generation in, or transmission near, where they live. Attempts by electricity utilities to site their infrastructure locally often result in eye-catching headlines such as 'CANCER FEARS OVER 'GREEN' POWER LINE. NEW PYLONS FOR RENEWABLE ENERGY' (*Sunday Herald*, Scotland, 27 November 2005).

Events in New York have played a very important role in the development of the power-lines-and-health story and it is worth quoting, in full, the objections of a New York State senator to a proposal to bring additional power to New York City.

STOP THE POWER LINE PROJECT

By William J Larkin Jr

I have joined four other state senators who represent an eight-county area in opposing the power line project proposed by New York Regional Interconnect Inc. The power line would cut through many communities in our districts stretching from Central New York through the Southern Tier to the Lower Hudson Valley.

New York Regional Interconnect Inc. wants to build a 1,200-megawatt transmission line that would run from Utica to Orange County. I strongly oppose the NYRI proposal. I have grave concerns about this project. Research shows there could be critical health issues associated with high-voltage electric lines. In addition, there are environmental issues as well as economic impacts. There is no question that these unsightly lines would affect the property value of area homes and the beauty and landscape of our region.

This is a billion-dollar project. It is being proposed not to meet the energy needs of upstate New York, but to serve only New York City, at the expense of upstate New York. I fully agree with my colleague, Sen. James Seward of Oneonta, who said, 'The power line means one thing: continued exploitation of upstate New York to serve downstate interests.'

The senator begins his list of objections with a mention of the possibility that there could be health issues associated with the proposed transmission lines. This question has also been taken up by British politicians, and a recent (2007) cross-party inquiry by a group of MPs, under the chairmanship of Dr Howard Stoate, has recommended restrictions on the construction of houses and schools near them.

Over the last 30 years or so, the supposed adverse health effects of the electric and magnetic fields (EMF) generated by power lines have received a great deal of publicity, but Senator Larkin's feeling that the electricity utilities were out to impose their unsightly infrastructure on rural communities for the benefit of remote city dwellers has a much longer history and, in recent years, it has added spice to the EMFs and power-lines debate.

The first British National Grid began operating in 1933 and its 132,000-volt (132kV) circuits were supported on pylons designed by the eminent architect Sir Reginald Blomfield. The grid had a robust defender in the Minister of Transport, Herbert Morrison, who answered protests over the erection of pylons on the South Downs with: 'They have a sense of majesty of their own and the cables stretching between them over the countryside gives one a sense of power, in the service of the people, marching over many miles of country.'

But the scheme did not find favour with the poet Stephen Spender, and his complaints still resonate today:

Now over these small hills, they have built the concrete
That trails black wire;
Pylons, those pillars
Bare like nude giant girls that have no secret.

Not many of the Central Electricity Board's engineers and members would have shared the poet's image and the 'pillars' were actually made of steel, not concrete, but 'they' and their pylons were the targets of Spender's complaint. The verse is also a reminder of just how much air quality has improved in recent years: he was writing soon after the lines had been built but the aluminium surfaces of the conductors were already blackened by air pollution.

EMFs and fright factors

Magnetic fields are generated whenever current flows. Exposure to them is ubiquitous and Earth's magnetic field has a magnitude of about 50 microtesla in the UK (a tesla is a unit of magnetic-flux density). This field is, of course, natural and relatively steady but the concerns described here relate to the artificial, alternating fields generated by the distribution and use of electricity.

The highly visible electric power network attracts attention but the background magnetic fields it generates are far from being the largest to which we are exposed. Beneath the high-voltage (400kV) transmission lines in the UK, the magnetic field will typically be about 10 microtesla, while that close to an electric razor will be about 2,000 microtesla. But, as we have seen, the pylons are often regarded as an imposition on the community and exposure to the magnetic and electric fields generated by the overhead wires they carry is then regarded as involuntary and different in nature from the exposures that result from the voluntary and usually short-term use of personal appliances.

Dr Jill Meara of the National Radiological Protection Board (now part of the Health Protection Agency) considered the rationality of risks and described how remote risks were overestimated and common risks underestimated. She pointed out the 'fright factors' that can apply to radiation, including:

- that exposure is involuntary;
- that radiation is inequitably distributed;
- that it is particularly dangerous to children or future generations;
- that it causes dreaded illness such as cancer; and
- the contradictory statements from authority.

The power-lines-and-health story that follows involves all of these factors and has evolved into a worldwide issue. There is now a vast literature on the subject – typing 'electric and magnetic fields + health effects' into a search engine will generate more than a quarter of a million hits. But, rather than cite references from the scientific journals, I will, wherever possible, link the story to material, often produced by NGOs, that can be downloaded free from the Internet. The story begins in Russia at the height of the Cold War.

From Russia – with a good deal of anxiety

Lenin famously said that 'Communism is Soviet power plus the electrification of the entire country' – and, with vast distances to cover, Russian engineers were active in the development of high-voltage power transmission. Notwithstanding the international tensions of the Cold War, technical exchanges were maintained between engineering colleagues from the East and West at the conferences of the International Council on Large Electric Systems (CIGRE) and, in 1972, they reported that maintenance staff in 500kV and 750kV switchyards were reporting a range of depressive symptoms that were related to their exposure to electric fields. These data were considered in a wide-ranging review of the epidemiological literature on EMF and health by the International

Commission for Non-ionizing Radiation Protection (ICNIRP). It reported,

> Concerns about possible psychiatric or psychological effects of EMF exposure were raised by investigators from the Soviet Union in the late 1960s and early 1970s on the basis of anecdotal reports of symptoms such as insomnia, memory loss and headache. However, these and other early reports have basically remained unconfirmed.

Nevertheless, the reports prompted research into the health of electrical-utility workers in the West and attracted the attention of Western objectors to the construction of new overhead lines.

Concerns about the possible health effects of power-frequency fields have been paralleled by similar ones about microwaves, radio and TV transmissions and, latterly, mobile phones. These concerns also have a Cold War, Russian element. Under the headline 'THE MICROWAVE FURORE', *Time* magazine of 22 May 1976 described how the Soviet Union had been beaming microwaves into the US Embassy to jam American monitoring equipment and, possibly, activate Russian bugs that had been planted inside the building. This was alleged to have been going on for some 15 years and staff who had worked in the embassy recalled strange ailments ranging from eye problems to heavy menstrual flows. Two former ambassadors had both died of cancer and all of these problems were, naturally, anecdotally related to the microwave exposure and helped to promote concerns about EMF in general.

The New York blackouts and their aftermath

The story now moves to America and the aftermath of the New York blackout of 1965, which gave rise to the urban myth that the power failure had been followed, nine months later, by a baby boom. Statistics did not support the 'baby boom' stories but the need for additional power was real enough and, in 1973, the New York State Public Service Commission received applications to build two additional 765kV transmission lines into New York. These proposals met with strong opposition from the regions through which the lines had to pass and the objectors were, by now, armed with the information from Russia that there may be health effects associated with the electric fields from high-voltage lines. Hearings started before administrative-law judges but the need for the power lines was spectacularly confirmed by a further and far more serious blackout of New York.

The New York power network was struck by lightning at 8.37 p.m. on 13 July 1977 and further lightning strikes together with equipment and operational failures resulted in a total blackout of the city and its airports by 9.27 p.m. People were trapped in lifts and electric trains and the city endured a night of looting, arson and rioting leading to thousands of arrests. Power was not fully restored until 10.39 p.m. the following day. It is not surprising that consent to build the power lines was granted by the New York Public Service Commission on 15 June 1978 but subject to the provisions that:

1. a research programme be established to determine the possible health risks arising from the electric and magnetic fields of overhead power transmission lines; and
2. a 350-foot-wide (106-metre) right-of-way corridor

surrounding each transmission line be established in which residence was not allowed.

On 7 February 1980, an agreement was signed between the New York Power Authority and the New York Public Service Commission providing for a $5 million research programme to be funded by the Power Authority and seven investor-owned utilities. This became known as the New York Power Lines Project and was supervised by the New York State Department of Health.

EMF research and the New York Power Lines Project

It would be human nature for the residents of New York State to start thinking, 'If "they" are spending all this money on research, there must be something in this.' As we have seen, the initial concerns related to the effects of electric fields on adults, but attention soon turned to the effects of magnetic fields on cancer in children. Nancy Wertheimer and Ed Leeper's study of childhood cancer in Denver, Colorado, was published in 1979, the year before the start of the New York Power Lines Project, but, as one of its many projects, a repeat of the Denver study was commissioned. This research was led by Dr David Savitz and, on 9 July 1987, the *New York Times* had the headline 'STUDY CITES INCIDENCE OF CANCER NEAR POWER LINES':

A study issued today by the New York State Health Department says that children with leukaemia or brain cancer are more likely than healthy children to be living in homes where the exposure to the magnetic fields generated by electric power lines is high.

The project, conducted in Denver, replicated an earlier study in that city that also found a correlation between the incidence of childhood cancer and the proximity of homes to the low-frequency magnetic fields produced by power lines. The new study shows the incidence of childhood cancer to be highest in homes closest to transmission wires designed to carry extremely high currents.

A study issued by an authoritative source, the New York State Health Department, had thus reported that exposures from heavy-current transmission lines (which can be seen as involuntary) were associated with the dreaded childhood diseases of leukaemia and brain cancer. Most of Dr Meara's fright factors were encapsulated in this report and growth in interest and concern about the issue had become inevitable.

Precaution becomes the policy

In June 1989, the *New Yorker* carried a series of three articles on the supposed perils of power-line fields by the journalist Paul Brodeur, who had campaigned earlier against microwave exposure. He was able to publicise his views on a number of TV programmes such as *Larry King Live*, and his articles formed the basis of his book *Currents of Death*, which was published in 1989 and followed by *The Great Power Line Cover-Up* in 1993.

The American government machine reacted to this evolving issue, and a report to the US Congress's Office of Technology Assessment in 1989 by Nair, Morgan and Florig produced a two-word sound bite, which is quoted to this day.

They suggested a policy of 'prudent avoidance' to address the EMF issue, which they defined as 'taking steps to keep people out of fields both by rerouting facilities and redesigning electrical systems and appliances'. Prudence was defined as 'undertaking only those avoidance activities which carry modest costs'.

Several utilities and their parent governments have enthusiastically taken up 'prudent avoidance'. It has been adopted in Australia, Sweden and several American states, including California, Colorado, New York, Hawaii, Ohio, Texas and Wisconsin. Other states have, however, rejected it because of a lack of evidence and scientific consensus. The proviso that only activities carrying 'modest costs' should be entertained rules out costly solutions such as putting power lines underground, and so 'prudent avoidance' policies are not likely to satisfy objectors who have fundamental objections to power-line development.

Science and the development of precautionary policies

The scientific programme of the New York Power Lines Project has been matched by similar efforts in the rest of the world. However, 30-odd years of well-funded research has produced a mass of literature but few definite conclusions. The studies in Denver by Nancy Wertheimer and David Savitz have been replicated and improved on in many other cities around the world. The World Health Organisation (WHO) has, over many years, run an 'EMF Project' and the results of the childhood-cancer studies are tabulated in its publication, 'ELF (Extremely Low Frequency) Environmental Health Criterion No. 238', which can be downloaded from the WHO website. The review of the epidemiological data on EMF and health, which was carried out for the ICNIRP, has already been mentioned in connection with the Russian microwave incident. The authors of that review included Anders Ahlbom, from the Karolinska Institute in Sweden, David Savitz, lead investigator of the Denver study, and Anthony Swerdlow, chairman of the UK Advisory Committee on Non-Ionising Radiation (AGNIR).They point out,

> Laboratory research has given no consistent evidence that EMF of the magnitude encountered in everyday life for a substantial period can affect biological processes or that EMF affects the risk of cancer in animals. The epidemiological literature is therefore particularly worth careful consideration because it is essentially on this evidence alone, at present, that suggestions about long-term effects on human health rest.

Their comments on the numbers of children involved in these studies are particularly important when it comes to the consideration of precautionary actions. When the data from six European countries, nine Midwestern and mid-Atlantic American states, five provinces of Canada and New Zealand were pooled, Anders Ahlbom found that children exposed to magnetic fields greater than 0.4 microtesla had their risk of leukaemia increased by a factor of two. But, in absolute terms, this enhanced risk applied to just 44 children out of the 3,203 who had been diagnosed with leukaemia. As the authors put it, 'Thus, fewer than 20 children among 3,203 with leukaemia represent the excess over expected numbers among children residing in homes with magnetic field exposure levels greater than 0.4 microtesla.'

The data relating EMF exposure and cancer have been evaluated by the International Agency for Research on Cancer (IARC), which is part of the World Health Organisation. Its monograph concluded that there is limited evidence for the carcinogenicity of extremely low-frequency magnetic fields in relation to childhood leukaemia. But, for all other cancers in relation to magnetic fields, the evidence was inadequate.

As a result, extremely low-frequency magnetic fields are classified as 'possibly carcinogenic to humans' (Group 2B, which is the same classification as coffee). Static electric and magnetic fields and extremely low-frequency electric fields are 'nonclassifiable as to their carcinogenicity to humans' (Group 3).

From the beginning, much of the research into EMF exposure and health has been funded by the electrical-power industry. As a consequence, there have been comparisons with the behaviour of the tobacco industry on smoking and health and criticism that the research was not 'independent'. This unjustified slur on the academic researchers involved has been accompanied by amused incredulity within the industry that there were those who could believe that they had the organising ability to run an international conspiracy.

Anyway, the Committee on Medical Aspects of Radiation in the Environment (COMARE) states on its website's home page that 'members have never been drawn from the Nuclear or Electrical Power Industries', so I will make use of its data, and the information to be found in its 11th report helps to put the EMF epidemiology into perspective.

Childhood cancer is, mercifully, a rare condition and, over the period of the COMARE data, the age-standardised rate (ASR) for childhood leukaemia averaged 37.7 cases per million children (aged 0–14) per year in England and Wales. The rate does, however, show variation around the country, ranging up to 48.3 (1.28 times the average) in Berkshire and 45.9 (1.22 times average) in Wiltshire.

At the start of this chapter, I quoted the Scottish *Sunday Herald*'s headline 'CANCER FEARS OVER 'GREEN' POWER LINE. NEW PYLONS FOR RENEWABLE ENERGY' as an example of the way communities can react to proposals to build lines of pylons. Journalists will almost invariably quote the relative risks when they are reporting epidemiological studies and avoid getting bogged down over their significance. In the article, one reads,

> The most disturbing evidence to date on the health hazards of pylons comes from a study by the Childhood Cancer Research Group at Oxford University. The study of 29,000 children in England and Wales found that those who lived within 200 metres of a high-voltage line had a 69 per cent increased risk of leukaemia. Those living 200 to 600 metres [219–656 yards] away had a 23 per cent increased risk.

When this extract is looked at in the light of the COMARE data, one could say that children living between 200 and 600 metres away from the power line seem to have the same risk of developing leukaemia as children in Buckinghamshire or Wiltshire and that would convey a very different message from that put over by the headline.

The study by the Childhood Cancer Research Group at Oxford University was a very large one and covered the children born in England and Wales between 1962 and 1995. The authors concluded that, if the higher risk in the vicinity of high-voltage lines really is a consequence of their proximity, then, out of the 400 to 420 cases of childhood leukaemia occurring annually, about five would be associated with the power lines. This highlights the very important question at the heart of any discussion about precautionary measures. If steps were taken to remove all children from the vicinity of overhead power lines, would one then be able to demonstrate that they had achieved their objective of reducing the incidence of leukaemia, given that the five annual cases said to be associated with the power lines is smaller than the year-on-year fluctuation in the total number of cases (400–420)?

The *British Medical Journal* accompanied its publication of the Childhood Cancer Research Group's study with a commentary entitled 'Power to confuse' by its science editor, Geoff Watts, which concluded,

Like the fluoridation of drinking water and the genetic modification of crops, the debate over power lines seems destined to be with us for a while yet. So, in these risk-averse times, and before activists begin blowing up pylons, a bit of perspective might help. In 2002, according to the Child Accident Prevention Trust, more than 36 000 children were hurt in road accidents and around 200 were killed. Another 32 died in house fires. Draper and colleagues reckon that five cases annually of childhood leukaemia may be associated with power lines.

In the UK, the National Radiological Protection Board (NRPB – now part of the Health Protection Agency) has the responsibility for advising the government on matters relating to standards of protection from radiation. The 2001 report of its Advisory Group on Non-ionising Radiation, produced under the chairmanship of the

late Sir Richard Doll, is available for downloading from the Health
Protection Agency's website. It reached the general conclusion:

> Laboratory experiments have provided no good evidence that
> extremely low frequency electromagnetic fields are capable of
> producing cancer, nor do human epidemiological studies
> suggest that they cause cancer in general. There is, however,
> some epidemiological evidence that prolonged exposure to
> higher levels of power frequency magnetic fields is associated
> with a small risk of leukaemia in children. In practice, such
> levels of exposure are seldom encountered by the general
> public in the UK. In the absence of clear evidence of a
> carcinogenic effect in adults, or of a plausible explanation from
> experiments on animals or isolated cells, the epidemiological
> evidence is currently not strong enough to justify a
> firm conclusion that such fields cause leukaemia in
> children. Unless, however, further research indicates that the
> finding is due to chance or some currently unrecognised
> artefact, the possibility remains that intense and prolonged
> exposures to magnetic fields can increase the risk of leukaemia
> in children.

The NRPB's 'Advice on limiting Exposure to Electromagnetic
Fields (0–300 GHz)' is based on the guidelines of the ICNIRP,
which take the view that the epidemiological data are 'insufficient
to provide a basis for ELF exposure guidelines'. The guidelines are,
therefore, designed to prevent effects on nervous-system function
and, for general public exposure to 50Hz magnetic fields, the Board
set a reference level of 100 microtesla – 250 times the 0.4

microtesla threshold from the childhood-cancer studies. However, the Board also noted,

> There remain concerns about possible effects of exposure of children to power frequency magnetic fields. The view of NRPB is that it is important to consider the possible need for further precautionary measures in respect of exposure of children to power frequency magnetic fields.

The government also established the Independent Expert Group on Mobile Phones under the chairmanship of Sir William Stewart. The group recommended a number of precautionary policies and, in particular, stated, 'In line with our precautionary approach, at this time, we believe that the widespread use of mobile phones by children for non-essential calls should be discouraged.'

The development of these precautionary policies is described by Adam Burgess in his 2004 book *Cellular Phones, Public Fears and a Culture of Precaution* (Cambridge University Press), and, by 2004, government was being recommended to consider precautionary policies for both mobile phones and power lines.

The formulation of regulations is a political rather than a technical matter. Costs have to be set against benefits and unintended consequences avoided if at all possible. The European Commission produced a directive that was designed to limit exposures to EMF in the workplace and was due to come into law in 2008. Following representation from the medical professions that the directive would severely limit the use of MRI (magnetic resonance imaging) scanners and increase exposures to ionising radiation because X-rays would have to be used instead, the

Commission has had to delay the implementation of the directive for four years for further investigation. The House of Commons Select Committee on Science and Technology's report into this fiasco makes interesting reading: 'Regulators will naturally wish to avoid anything similar in the future.'

The Department of Health reacted to the recommendation that further precautionary measures should be considered by the formation of the Stakeholder Advisory Group on Extremely Low-Frequency EMF (SAGE), with the remit to 'bring together the range of stakeholders and explore the implications for a precautionary approach to ELF EMF and make practical recommendations for precautionary measures'. The IARC classification that EMFs were a 'possible carcinogen' was based on the childhood-cancer data, and the WHO, in its memorandum on ELF EMF, takes a severely limited view of precautionary action.

Provided that health, social and economic benefits of electric power are not compromised, implementing very low-cost precautionary measures is reasonable and warranted.

However, a review carried out for the California Public Utilities Commission in 2002 concluded that the EMFs can cause a much wider range of diseases, including adult brain cancer, Lou Gehrig's disease and miscarriage, in addition to childhood leukaemia.

Naturally, if one favours the 'California' interpretation of the science, then one will believe that limiting the population's exposure to electric and magnetic fields will save a larger number of people from disease than would be predicted if one favoured the WHO/Health Protection Agency view that the risk is limited to childhood leukaemia.

The SAGE report shows that the 'stakeholders' were divided on this issue and the economic implications of the two interpretations as set out on the report's Page 51 merit repetition:

- cost per home removed from a field of 0.4 microtesla: £20k–£160k;
- health benefit per home removed from a field of 0.4 microtesla on the WHO/HPA view of the science and assuming magnetic fields above 0.4 microtesla do cause leukaemia: £1k;
- health benefit per home removed from a field of 0.4 microtesla on the 'California' view of the science perhaps a hundred or so times larger.

On the following page, one reads:

> Stakeholders take different views on the 'WHO/HPA' and 'California' views of the science and have consequently not been able to reach a consensus on the advice that stems from these views. Therefore SAGE needs to set out alternative advice to Government depending on which of these views is followed through, recognising that not all stakeholders are comfortable with each piece of advice.

The minority reports at the end of the document show just how uncomfortable some of them were.

SAGE had been considering the option of stopping the building of houses and schools within specified distances (60 metres/65 yards in the case of 400kV) of overhead power lines and of stopping the building of new power lines close to existing buildings. This is likely to have the unintended consequence of making existing houses less attractive. As the report says, 'However, if action were taken to prevent new instances of lines and homes in proximity while allowing the existing instances to continue, this is likely to create some concern among the people affected.'

Prior to the report's publication and posting on the Web, the *Daily Telegraph* of 1 May 2006 reacted to a leaked version with the headline 'PYLON CANCER FEARS PUT £7BN BLIGHT ON HOUSE PRICES': 'Up to £7 billion will be wiped off property values if the government accepts the advice of experts that homes should no longer be built near overhead power lines because of possible links with childhood leukaemia.'

The SAGE report was published in April 2007 but the Department of Health has not yet, at the time of writing, given its formal response. However, it had asked for advice from the report from the Health Protection Agency.

The NRPB, which is now part of the HPA, had started the hare running with its suggestion that the government should consider the 'possible need for further precautionary measures' and, as one would expect, a letter from Professor Pat Troop, the HPA chief executive, reconfirmed the NRPB scientific position. In particular, she confirmed that 'the evidence for an association between exposure to EMFs and a number of other diseases (the California position) is much weaker than that for childhood leukaemia and also lacks plausible biological support'.

Given this position, the 'corridor option' for separating new buildings from high-voltage power lines is not supported by the cost–benefit analysis, even if one assumes a causal relation between EMF exposure and childhood leukaemia'.

Very large sums of money have been devoted to EMF issue and

she notes that the costs of any proposed precautionary approaches should be considered alongside other potential uses for the money, for example, to improve services for the treatment of leukaemia, or to enhance research into its causes and treatment.

Meanwhile, the media have not lost their appetite for EMF stories, on the mobile-phone front. Instance the *Guardian*'s headline of 25 July 2008, on an inside page: 'LIMIT MOBILE PHONE USE, CANCER EXPERT TELLS STAFF'. On the electrical-power front, we are often reminded of the need for alternative energy and the overhead wires needed to connect the proposed wind farms and other forms of 'green' energy are likely to attract local opposition. Mothers with small children will continue, quite naturally, to demand confirmation that power lines near their houses are 'totally safe' and the scientists will continue to have to say that no research programme can ever give such a reassurance. But, as New York showed in 1977, when the lights go out, modern society can face real and truly terrifying risks.

3

GRIDLOCKED BRITAIN: A TRANSPORT POLICY?

INTRODUCTION

STANLEY FELDMAN

IN JULY 2008, it was officially confirmed that the vehicular transit time across London was longer than before the introduction of congestion charging. Although the number of private cars crossing London has decreased, and millions of pounds have been spent, it now takes longer to get from one side of the city to the other than it did in the days of the horse and carriage.

The same story has been repeated all over the country. Billions of pounds have been spent in the last ten years on a transport policy dictated by political rather than economic considerations. As a result, journey times have got longer and congestion has worsened. The prospect of 'gridlocked Britain' has led the government to float suggestions of the Draconian measures that may be necessary to try to stop people from using the roads. The billions of pounds that have been spent on the railways and public transport has not solved the problem. Unless there is a totally new approach to the problem it will get worse as the population of this crowded island increases.

Congestion charges have not been the answer. Indeed, any initial benefit has been offset by the congestion caused by the increase in the number of buses. The problem is particularly acute in London where Transport for London (TfL) has led a campaign to reduce the number of private cars on the roads.

The enormous increase in the number of traffic lights has caused longer transit times, higher fuel consumption and confusion for the motorist confronted with multiple sets of traffic lights. It has had only a marginal effect on the number of road casualties.

The reduction in road width for central pedestrian islands, bus lanes and cycle paths has made it inevitable that roads will become blocked when a lorry parks to load or unload. The creation of turning-left lanes before traffic lights and at roundabouts has created the absurd situation where at a point of congestion, just before a traffic light, there will be a long queue for those going ahead while the turning-left lane is empty.

As any A&E department will confirm, cycle accidents are frequent and increasing. How can the body that introduces multiple traffic lights to reduce road accidents encourage cycling, which results in more accidents?

It is clear that the transport policy over the past ten to twenty years has been a dismal failure. The prospect of gridlocked Britain is real. It is time to reconsider the problem in the light of economic necessities rather than political priorities.

16

A POLICY FOR TRANSPORT?

PAUL WITHRINGTON

DOGMA
Car use should be curtailed and people encouraged
to travel by bus and train.

The background

The ability to travel was once the sole privilege of the wealthy. The ready availability of cheap transportation has resulted in one of the great changes in social behaviour over the past hundred years. Previously, most people lived their entire lives within a few miles of their birthplaces. The mould was broken initially by the train, the bicycle and the bus. Since World War II, the motorcar and aeroplane have led to the unprecedented sense of liberation that we enjoy today.

However, the car has brought obvious problems. Congestion is said to be costing the nation £15–20 billion annually, of which £7.5–8 billion is in London. Over the decades, thousands of people have been killed and injured in road accidents. Those without access to a car have been disadvantaged. It has contributed to atmospheric pollution and it has been portrayed by the environmental lobby as a major contributor to carbon emissions and global warming

Because of the scale of those problems, the car has been demonised in recent years. It has been seen as a cause of problems, rather than as a public benefactor. It has led to a determined campaign against the car, which is both misplaced and impossible to win.

This antipathy to road transport is supported by the sentimental attachment that the nation has for the railway, a kind of religion, beyond criticism or rational debate. Any shortcoming or failure of the railways is ignored or attributed to human frailty.

As a result of this prejudice against the car, we lack a sensible transport policy. Instead, there has been massive public expenditure on policies that, at the national level, are almost entirely ineffective.

The 10-Year Plan

In 2000, the government published 'Transport 2010: The 10-Year Plan'. The target was to increase rail use by 50 per cent and bus use by 10 per cent. We were told that it would greatly reduce road congestion. It was supported by, if not the brainchild of, the Commission for Integrated Transport

However, that policy did not take account of the obvious fact that the bus and train each accounted for only 6 per cent of the passenger-miles travelled. Even if the targets could be met, there would be little overall impact on the increase in car use, which was growing at a much faster rate. Although it has in fact increased by 10 per cent over the period (Figure 16.1 illustrates the point), it is difficult to discern the difference in car use as a result of this policy.

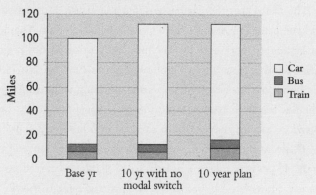

Figure 16.1: Passenger-miles expressed as percentage of the base year

If that is not enough to illustrate the limited impact of the plan, consider the history over the past 50 years. In 1955, the bus and train (including London Underground) accounted for about 60 per cent of the passenger-miles travelled. Today, bus and rail travel together account for a total of 12 per cent of passenger-miles, while 85 per cent of passenger-miles are made in cars. Passenger-miles by car have increased sevenfold. Figure 16.2 illustrates the overwhelming size of this change.

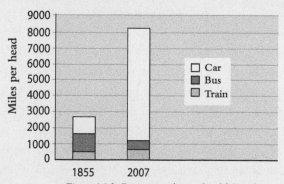

Figure 16.2: Passenger-miles per head by mode

The fact is that the car has enabled a land-use distribution and lifestyle that is almost impossible to serve by bus or train. That leaves the bus and train to serve the declining proportion of jobs and retail businesses that remain rooted to town centres. Nevertheless, billions of pounds have been spent, and are continuing to be allocated, to schemes based on the presumption that congestion really can be solved by getting people out of cars and into buses and trains.

Those wishing to claim success for the 10-Year Plan will point to the 40 per cent increase in train usage that has occurred over the decade. However, the percentage of passenger-miles travelled on national rail remains close to 6 per cent of the total miles travelled. Since rail serves destinations that are difficult to reach by car and the car serves journeys impossible to make by rail, it is probable that relatively few people outside London have voluntarily transferred from one mode to the other. As a result, at great cost to the nation, we have encouraged people, statistically those in higher income brackets, to commute relatively long distances by rail, establishing lifestyles that can be sustained only by a massive public subsidy.

Public transport

Despite these measures, outside London there has actually been a 10 per cent decline in the use of public transport, despite the 10-Year Plan's wish for a 10 per cent increase. This is because of rising car ownership and a change in the patterns of employment and retailing, which have deserted town centres in favour of out-of-town locations that cannot be served effectively by a bus. Ironically, this dispersal of land use has been encouraged by well-meaning policies, such as limiting car parking in town centres and, where it exists, charging excessively for it. Congestion charging, if it becomes widespread, may very well accelerate the trend and result in a consequential decrease in the use of public transport.

The reality is that the car is the second most expensive purchase any family makes. It is the principal and preferred way of meeting the travel requirements of the majority of the population. The bus and train are major players in the transport systems only of large towns. Even in London, once one moves outside the centre, the car accounts for 70 per cent of motorised trips and possibly 80 per cent of passenger-miles travelled.

Even on journeys where bus or rail may appear to be an obvious choice, there are significant disincentives to their use. Firstly, there is the journey to and from the bus stop or rail station. Secondly, few people can programme their lives to suit bus and railway timetables; consequently, time is lost waiting. Thirdly, the train and the bus are often unsuitable for those with heavy shopping bags, the disabled, the elderly and mothers with babies.

It is not surprising that, with the exception of central London, public transport, although vital to some sections of the community and for some journeys, plays no more than a trivial role in strategic terms and can never achieve more unless there is a prohibition of the way of life that the car has enabled.

Even if that happened, we doubt that there would be a large-scale shift to the bus. Instead, there would be a decline in out-of-town industrial estates and shopping centres. It would cause a return to local employment and shopping. It would return us to the 1950s.

Notwithstanding these facts, the policy over the past several years has been focused almost entirely on 'getting people out of cars' and on a drive to improve road safety. This policy has been pursued at the expense of the need to keep the traffic moving and the stated objective of reducing carbon emissions.

Overlying the whole is the pervasive idea that streets are used only by pedestrians. The fact that people and businesses need motor cars, buses and lorries seems to have escaped notice.

What have they done?

There has been an extraordinary growth in the number of road junctions controlled by traffic lights, in the number of signal-controlled pedestrian crossings and in road markings at junctions that allocate a specific lane for each and every turn that has to be made. There has also been a campaign against speed, by the use of speed cameras, speed humps and ever-lower speed limits. Additionally, the tax on cars has been raised and now reflects their fuel economy.

These changes have had a major impact on the lives of ordinary folk. Capacity at the most critical points in the network has been reduced. There is now congestion where none need exist. Signal-controlled crossings bring traffic to a standstill, often on major traffic routes, long after a lone pedestrian has crossed, creating yet more congestion. The road markings at roundabouts and junctions lead to massive queues for major movements and empty lanes for minor ones.

The speed-camera campaign has led to the prosecution of millions, with tens of thousands losing their licences annually. Some of the latter will have lost their livelihoods; others will have continued to drive illegally and with no insurance. Most will have been penalised, although they were driving perfectly sensibly for the conditions. Consequently, many will harbour resentment against the authorities and the police.

We do not have a good estimate of the total cost of making these changes, but a signalised road junction may cost up to £500,000, traffic lights at pedestrian crossings in excess of £100,000 each and a set of speed humps £1,000–£3,000. We believe the total cost of those measures may exceed half a billion pounds. The consequential congestion costs imposed on the travelling public may dwarf that cost.

The other major policy impact has been a massive subsidy to the railways. This is likely to amount to £100 billion for the 20 years to 2015 – equivalent to £4,000 for every household in the land.

Has there been anything to cheer about?

Congestion is undeniably much worse than ten years ago. Except in London, bus use has declined by 10 per cent over the last decade, although rail use has soared by 40 per cent. Those who need large cars because of family size, or other reasons, are penalised, while the truly rich, who may buy gas guzzlers for sport, will be less affected. Worse still, the previous decline in casualties in road accidents appears to have been sabotaged.

The 40 per cent rise in rail use could be a cause for cheer in that dismal catalogue; however, it has also given cause for concern. Our reasons for saying so are given in the next chapter. The astonishing growth that has been achieved has been at immense cost. It turns out not to be environmentally sound, benefits the rich, not the poor, encourages lifestyles that can be maintained only by further massive subsidy, and has done little to reduce road congestion.

Speed on the roads

The punitive nature of the speed-camera regime and its wide-ranging effect on the population at large make the attack on speed particularly worthy of comment. The regime became effective from the year 2000.

Theoretical research published in *Mathematics Today* by Rose Baker, Professor of Statistics at Salford University, suggests that the average driver will face a ban once every 15 years. One offence in two years is the average but the variation means a quarter may get banned every seven years, by pure chance, while nine per cent escape altogether.

The practical effect of the cameras has been the prosecution of millions, mostly those who were driving sensibly for the conditions, the loss of licences for perhaps hundreds of thousands and, for many, the loss of employment. A side effect may be that many more people than previously are now driving illegally and hence uninsured.

The penalties would perhaps be worthwhile if it could be shown that speeding (narrowly defined as breaking the speed limit) were a significant cause of road deaths, or, more importantly, that there had been a marked acceleration in the established rate of decline in road deaths and other injuries. However, as we shall see, neither of those suppositions is true.

Firstly, the Transport Research Laboratory's report LR 323 published in 1998 provides that in only 7.3 per cent of accidents was 'excessive speed' recorded as a contributory factor. In confirmation of that low number, in 2005, and according to data in Table 2 of the Department for Transport's report with the title 'Contributory Factors to Road Accidents', 'speeding' accounted for only 5 per cent of all recorded causes for accidents where there was a fatality, for only 3 per cent in accidents where the most serious injury was designated as 'serious', and for only 1.8 per cent in accidents where there were only slight injuries.

Despite that, the Government and the Department for Transport (DfT) have suggested that 30 per cent of accidents are due to 'speed' or 'excessive speed'. The authorities have then allowed those terms to be confused with the term 'speeding', so creating the quite wrong impression that breaking the speed limit is a major cause of road accidents.

Secondly, the annual rate of decline in deaths per vehicle-km for the 30 years to the year 2000 averaged 5.3 per cent but, instead of that beneficial trend accelerating under the impact of present policies, the rate of decline collapsed to 3.4 per cent. Indeed, had the historic trend continued, in 2007, instead of 2946 deaths there would have been 2544, some 400 fewer than actually occurred.

Against that background, it is difficult to see how those who defend present road-safety policies can claim that those policies have saved lives. After all, the data suggests precisely the reverse. In contrast to the collapse in the downward trend in deaths, the decline in the reported seriously injured casualties appears to have held up. However, a consequence is that the ratio of the reported seriously injured to the killed has declined from 13 to 9.5 or by 26 per cent. That is difficult to explain since, if anything, one might have expected modern medicine to save more lives than in the past, so increasing the ratio. The only explanation for this anomaly that seems likely is that, under the impact of targets, there is now more reluctance to classify an injury as serious than in the past. In any event, perhaps driven by different targets, the *British Medical Journal* of June 2006 reported an increase of 1 per cent in road-traffic casualties, not a decrease.

In addition, we have the Transport Research Laboratory reports LR 421 and 511. These set out to find a relationship between speed and accidents. Since reducing speed to zero would also reduce accidents to zero, it was scarcely surprising that a relationship was indeed found. That relationship is often summarised as a 1 mph reduction in speed leading to a 5 per cent reduction in accidents, although the number varies according to the type of road. We do not believe this finding can usefully inform policy. Nevertheless, it has been used to justify speed limits so low that the law is being brought into contempt.

There may be three reasons for the failure of these-policies. The first is the distraction of the constant threat of being caught 'speeding'. Instead of concentrating on the road ahead and making sensible judgements, drivers may now be concentrating on their speedometers. The second falls within the slogan 'Treat people like idiots and, surprise, surprise, they will behave like idiots'. The motorist is now expected to 'drive by numbers' both with regard to speed and to his use of road space, particularly at junctions. It has eroded driver responsibility and consequently sabotaged the development of mature behaviour. Thirdly, there has been a reduction in the number of traffic police.

Against that background, instead of the punitive approach of the past few years, we advocate one based on education, designed to develop a polite and considerate driving population. Traffic police should also be present in greater numbers but with the discretion as to when to prosecute.

Sustainable and integrated?

The anti-car policies are supported by the vague notions of 'sustainability' and 'integration'. That has serious side effects. On the strength of this jargon, consultants for large-scale developments have gleefully promised bicycle networks, orbital bus routes, park-and-ride, etc., and, hey presto!, they have halved the forecasts for the rise in traffic. The planning authorities then concur with the apparently ludicrous conclusions, believing them to be 'eco-friendly'.

That would be all very well if there were a shred of evidence that any of those policies could have more than a trivial effect. Our view is that there is no such evidence. Instead, for political convenience or for fear of being overridden by a secretary of state determined to force through a massive house-building programme, the planning authorities are nodding through major developments that are likely to cause massive congestion in the future

Scheme evaluation

The economic analyses now being carried out, as cost-benefit studies, for major public-transport proposals are based on a false premise. The consequence is that billions of pounds of taxpayers' money are being diverted to schemes that can never pay for themselves, even when a cash value is assigned to the time savings, etc. that are supposed to arise.

Cost–benefit analysis was originally developed for the evaluation of road schemes where, because there were no direct payments by the users, the cash value of the time savings, the accident savings and reduced vehicle operating costs were compared with the construction costs.

When the same process is applied to public transport, the benefits are inflated by the 'incremental' fares. These are the fares that the proposed programme would generate but that go to the operator. Our view is that that is quite wrong. All such payments are transfers where the loss to the one party (the customer) balances the gain to the other (the operator). Our illustration relates to Crossrail. However, the same applies to all major rapid-transit schemes and to the railways generally.

For Crossrail, the lifetime value of the fares revenue is some £13 billion. That includes some £7 billion, which would have gone to existing rail services, leaving incremental fares generated of around £6 billion. That huge sum was subtracted from the lifetime costs of £13.7 billion to provide £7.7 billion. Loss of indirect tax at £1.2 billion was then added to yield a net cost of about £9 billion. It is that net cost that has been compared with the lifetime user benefits of £16 billion, giving an apparently satisfactory cost–benefit ratio.

However, an entrepreneur must take account of tax when deciding on a major investment. In a cost–benefit analysis, such a transfer leaves total resources unchanged. Indeed, rather than the loss of tax to the government being a cost, it is perhaps a benefit to the nation, in that government seldom spends money as well as the private sector does. In any case, the government could, at the stroke of a pen, make good any tax loss.

The classical argument is that the cash lost to the passengers in paying for tickets clearly equals the gain to the service provider, leading to zero net benefits attributable to the transfer.

To illustrate the point: suppose London's transport operator ran taxis and buses as well as public transport. The incremental fares for Crossrail would fall to perhaps £5 billion because revenue previously taken by the buses and taxis would be lost. Going a step further, suppose the operator also ran restaurants, retail malls, etc. Incremental fares might then fall by another couple of billion. Taking an extreme case, if the transport operator ran the entire economy, then incremental fares would fall to zero, illustrating the point that fares, whether incremental or not, are a transfer payment of no social benefit in relation to cost.

What has happened is that the accountants have ignored the fact that incremental fares are at the expense of other sectors of the economy. In short, they have muddled financial analysis with cost–benefit considerations. That has vastly (and erroneously) increased the benefit to the cost ratios of many public-transport proposals.

Instead of the fatally flawed cost–benefit approach, we would advocate a financial analysis. Such an analysis would reject projects where whole-life costs are greater than the sum of the whole-life fares plus the associated changes in land values. On that basis, Crossrail should not be built unless it is likely to yield an increase in land values greater than £7.7 billion (the gap between net fares and costs).

Any other approach will leave the taxpayer paying for fairy gold for ever and ever, enabling those who rule us to extract ever higher taxes so as to pay for ever more glorious schemes, none of which the market would ever support.

Climate change and emissions

We believe this is a red herring. It is nevertheless being used as an excuse for using cars as a means of raising taxes.

The truth is that, if we wished to reduce the CO_2 emissions of motor traffic, then, rather than increase road tax on large cars, we would abolish the tax along with VAT on motor sales and transfer the cost burden to fuel. The increase in fuel costs would go some way to mimicking congestion charging. It would also encourage the purchase of efficient vehicles and limit their use. High fixed taxes may actually be encouraging car use, as the higher the mileage travelled the lower the cost per mile driven.

While conserving energy is sensible, policies designed to encourage one mode of transport rather than another should be resisted, or at least postponed until there is better scientific evidence to justify their introduction as the evidence does not necessarily favour rail transport.

ROAD VERSUS RAIL

PAUL WITHRINGTON

DOGMA
Rail is good, road is bad.

STEWART JOY, THE CHIEF economist to British Railways in the 1960s, wrote in his book *The Train that Ran Away** that 'there were those in the British Transport Commission and the railways who were prepared, cynically, to accept the rewards of high office in return for the unpalatable task of tricking the Government on a mammoth scale. Those men', Joy wrote, 'were either fools or knaves.'

Then, as now, railway propaganda has perpetuated the belief that rail is far safer than road transport ever can be, has by far the higher capacity, is intrinsically 'green', is in some way vital to the nation, benefits the poor and brings development in its wake. Examining the evidence suggests that none of that is true. Instead, the gap between the myth and reality is overwhelming.

*first published 1973 (Ian Allen Ltd)

Safety

'The Railways are efficient as killers': in a paper in the *Journal of the Institution of Highway Engineers* (1997), J J Leeming, estimated that, per hour spent on them, 'the railways are twice as dangerous as the roads'.

In contrast, the report of the Transport Committee's inquiry into the 'Future of the Railway' in April 2004, ordered by the House of Commons, states that,

> The figures comparing road and rail fatalities are telling. In 2002, 3,431 people were killed on the roads while according to the HSE (Health and Safety Executive) 50 passengers, railway staff and other members of the public were fatally injured and 256 people died as a result of trespass and suicide on the railway.

The committee went on to write, 'The Strategic Rail Authority (SRA) points out that "on average more road users die in accidents each day than rail passengers in a year".'

Those numbers were inspired by the railway industry. They are correct but grossly misleading. First, there are currently 17 times more passenger-miles travelled by road than rail. Consequently, the numbers quoted exaggerate in favour of rail by a multiplier of 17. Second, the statement by the SRA compares passengers killed in so-called train accidents (and, if you fall out of a train, that is not a train accident: it is your fault) with all those killed on all roads, including pedestrians, cyclists and people on motorbikes. That introduces a further and similarly large exaggeration in favour of rail.

The railway lobby also likes to boast that 'no passenger died in a train accident last year', whenever there is a year when that happens. What the railway lobby does not point out is that, since 1915, 1,370 people have died in train accidents – an average of 15 per year. Furthermore, before the phasing out of slam doors, more people died from falling out of trains than in train accidents.

Taken together, these propaganda items have created the impression that rail is indeed overwhelmingly safe compared with road, when nothing could be further from the truth.

If we are to make useful comparisons, we must compare the casualties per mile travelled and consider the two quite separate constituencies:

- passengers, who wish to emerge from the system unharmed; and
- the population as a whole (in addition to passengers, that constituency includes staff, postal workers and people on railway business, none of whom are classed as 'passengers', and crucially trespassers, but probably not suicides).

It is also essential to compare similar modes of transport. Hence, we compared deaths of passengers per passenger-mile by train with the same for express coaches operating on the comparable strategic road network. Since the annual variations in the small number of deaths per year by these modes of transport make a single year's comparison statistically unreliable, we have used the rates for the ten-year period 1996–2005. For that period, we found that the deaths per passenger-mile by rail were nearly double those by express coach, with the caveat that in calculations we had to assume an average coach occupancy. We chose 16, which is midway between the reputed occupancy of 25 for a coach leaving Victoria Coach Station and the average of 9 for all buses.

We also found that, system-wide, including trespassers but not suicides, the deaths per passenger-mile by rail for the ten-year period was 50 per cent above that for the motorway and trunk-road system for the single year 2002. We then excluded deaths of cyclists, pedestrians and people on motorbikes from the road statistic on the grounds that, if the railways were paved, such people would be excluded from the rights of way. That produced a death rate that is half that of rail.

If, rather than deaths, the costs of killed and seriously injured (KSI) casualties are considered, the advantage in favour of road transport widens. For example, (a) the KSI casualty cost per passenger-mile associated with passengers using trains is three times higher than for passengers using express coaches; (b) the system-wide KSI casualty cost per passenger-mile by rail is 2.1 times that for the strategic road network if pedestrians, cyclists and people on motorbikes are included and 2.7 times higher if those classes of people are excluded.

Lastly, deaths to passengers in train accidents, the statistic most often published by the railway lobby, are a trivial measure of the problem as a whole in that their cash value amounted to only:

- 10 per cent of the cost associated with KSI passengers injured in 'train' and 'movement' accidents;
- 5 per cent of the cost of the KSI passengers injured in the envelope bounded by the ticket barriers; and
- 1.5 per cent of all the KSI casualties (excluding suicides) injured within the railway system.

Capacity

Bombardier, the train-carriage-manufacturing company, in its evidence to the Transport Committee on the future of the railway claimed that to carry 50,000 people it would need:

- a 175m-wide (574ft) road if car transport was used;
- a 35m-wide (115ft) road if used by buses or coaches; and
- a 9m (29ft) track bed for a metro or commuter railway.

We presume Bombardier based its figures on the capacities of streets in towns instead of motor roads free of congestion. Whatever the case, in contrast to Bombardier's assertions, a report by the US Department of Transportation, in January 1973, found that, 'A single lane of highway, used exclusively by buses, would provide a passenger-carrying capacity in excess of almost any known level of demand.'

We have three illustrations that contradict the Bombardier statement.

First, on the approaches to central London terminals in the peak hour, the surface rail network carries an average of only 10,000 passengers per inbound track. Those passengers will be travelling in crushed conditions. All of the 10,000 would find seats in 150 coaches, each with 75 seats. Those coaches would fill one-seventh of the capacity of one lane of a motor road, the same width as required by a train. Hence, it is clear that, if the network were to be paved, then, even in the peak hour, the express coaches replacing the trains would occupy only a small fraction of the capacity available.

Secondly, Welwyn viaduct, a notorious bottleneck on the East Coast Main Line, carries 14 London-bound trains in the peak hour.

If each train has ten carriages, then the flow will amount to 140 carriages per hour. That is equivalent to about 200 50-seat express coaches – sufficient to occupy one-fifth of the capacity of one lane of a motor road.

Thirdly, there is the New York contraflow bus lane, which serves the main bus terminal via the Lincoln tunnel. The lane is some 3.3 metres (11ft) wide and 6.4km (4 miles) long (including 2.4km/1.5 miles in tunnel). In the peak hour, it carries *circa* 700 45-seat coaches, which provide more than 30,000 seats. The trains serving Victoria Main Line also carry *circa* 30,000 passengers in the peak hour. However, in contrast to the passengers in the New York bus lane, the rail passengers are in crushed conditions, often standing, and the trains require *four* (inbound) tracks instead of one. The difference between the railway lobby's propaganda and reality with regard to the key factors of safety and capacity is overwhelming. Consequently, people who have been subjected to that propaganda tend to feel that we have played some sort of 'trick'. However, there is no trick. Instead, in all ways, except the romantic, the gap between railway propaganda and reality is so large as to beggar belief.

Productivity

Passenger and freight flows are measures of productivity. We calculated the network-wide averages for Network Rail and for the strategic road network by dividing estimates of the passenger-miles and tonne-miles by the track and lane lengths. That showed that, per mile of lane, the motorway and trunk road system is used 2.5–3 times as productively as the tracks on the national rail network, despite the latter having the advantage of serving the hearts of our towns and cities. Figure 16.3 illustrates the point.

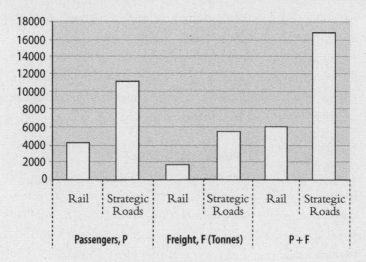

Figure 16.3: Average daily flows per track and lane in 2007

An alternative calculation that produces an enlightening result is to convert the passenger and freight flows by rail to equivalent express coaches and lorries. If the coaches had as few as 20 people aboard and if the rail freight were in lorries, carrying 30 tonnes outbound and coming back empty (providing an average load of 15 tonnes), the average flow across the network would amount to a trivial 330 vehicles per day per track, a flow so small that it would be quite lost on a motor road.

One reason why rail freight in the UK is unattractive is that the average line haul is only some 130 miles long. Hence, the cost of dragging freight by lorry to and from the rail heads adds greatly to the costs. In comparison, a freight train in the USA may be a mile long and the line haul 2,000 miles long. It is in those circumstances that the rail freight may be economical but at considerable environmental cost – some of the marshalling yards are over 15 miles long. Local distribution from the marshalling yards, typically up to about 500 miles, is by lorry.

Numbers apart, one has only to observe the use of a railway next to a motorway or trunk road for it to be obvious that the railway is a waste of space. Trains use the track intermittently while the flow of traffic on the roads is constant.

Vital to the nation?

The chairman and chief executive of the SRA, Richard Bowker trumpeted the fact that 'nearly half the population uses a train at least once a year' (see the Foreword to his *Everyone's Railway: the wider case for rail*, 2003). However, the corollary, namely, that half the nation uses a train less than once a year, seems to have escaped his notice.

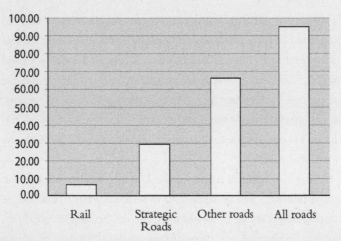

Figure 16.4: Passenger-miles per cent by rail and road

Figure 16.4 illustrates the trivial contribution (6 per cent of passenger-miles, and fewer than 2 per cent of passenger journeys) that rail makes to the nation's needs. Further, since 70 per cent of all passenger rail journeys are in London or the Southeast, for most of the country, these numbers exaggerate the contribution of rail. National Travel Survey data shows that the car is the dominant mode of transportation even for the longest journeys. It carries 72 per cent of journeys with lengths in the range of about 250–350 miles leaving 8 per cent to the express coach, 14 per cent to rail and 5 per cent to air. For journeys more 350 miles long, 42 per cent are by car, 5 per cent by express coach, 39 per cent by air and only 12 per cent by rail.

The same picture emerges when we consider freight. Four per cent is carried by pipeline, 8 per cent by rail, 20 per cent by water (mainly inshore shipping) and 68 per cent by road. If road and rail transport alone are considered, then we have 12 per cent carried by rail and 88 per cent by road (55 per cent is on strategic roads, leaving 33 per cent to other roads). Figures 16.4 and 16.5 illustrate those numbers.

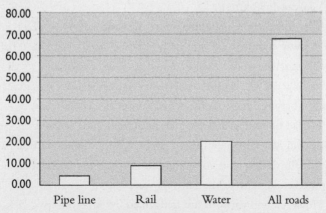

Figure 16.5: Tonne-miles: per cent by mode (2007)

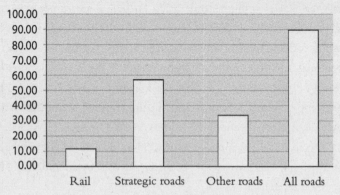

Figure 16.6: Tonne-miles: per cent road and rail (2007)

Fuel and emissions

F A S Wood, chairman of the National Bus Company, commenting on a British Rail advertisement, which claimed that a train conveying 250 passengers from London to Newcastle uses only 400 gallons of fuel, wrote in a letter to *The Times* of 15 April 1974 that 'six of our long-distance coaches will do the same job using about 150 gallons [682 litres]'.

The comparison made between road and rail by the DfT and the railway industry are almost pointless. That is because (a) the operating conditions on the roads are not comparable to those enjoyed by rail and (b) there really is no possibility of any significant transfer between the two modes of transport.

Instead of comparing the fuel consumption of trains with cars, buses and lorries operating in average conditions, we compare trains with road vehicles operating in uncongested conditions such as they would enjoy if the railways were paved.

However, comparisons are bedevilled by an astonishing lack of robust data. For example, *Transport Statistics for Great Britain*

(TSGB) for 2007 gives a figure for the energy used by rail in 2005 that is 18 per cent below that reported for the same year in the 2006 edition. Furthermore, there is no division between freight and passenger services.

For rail passengers, the best data source is probably the Rail Safety and Standards Board's (RSSB) report 'Traction Energy Metrics', published in 2007. Those data relate to the year 2005. They are probably robust since they were provided by the Association of Train Operators. The same publication purports to provide data for rail freight. However, it used figures from Transport Watch (of which I am director) as the source. That data came from Network Rail staff and relates to the year 2002–03. Subsequently, we have come to distrust the source because the passenger data provided at the same time led to a fuel economy that was at variance with both that from the RSSB report and others that we had from British Railways relating to 1990. In addition, Network Rail said that the diesel consumption should be split approximately 50/50 between freight and passengers, whereas the RSSB data provide a quite different division.

Hence, for freight, we have used data relating to 1998 provided to us by Tim Murrells of NETCEN*, citing the National Atmospheric Emissions Inventory on behalf of the DETR**. We find it astonishing that this appears to be the best information available to the nation, yet the literature is full of apparently authoritative statements about the relative carbon emissions of rail.

In our calculations, we assigned the transmission and generation

*National Environment Technology Centre
**Department for Environment, Transport and the Regions, renamed Department for Transport (DfT)

losses of electricity (amounting to 17 per cent of the total generated) proportionally to the end user, in this case rail. Similarly, we have increased the nominal fuel consumption of diesel-powered vehicles by 10 per cent to allow for losses in refineries (5.7 per cent) and in transporting the fuel to filling stations. That differs from most other sources, where either no such allowances are made or transmission losses alone are allowed for. In deference to the UK vernacular, we express the fuel consumption in terms of passenger-miles per gallon.

On that basis, we found passenger rail returned 92 passenger-miles per gallon. In comparison, an express coach operating on a railway alignment may be expected to achieve 10 miles per gallon. With 20 people aboard, and assuming 10 per cent losses at refineries and in transporting the fuel, that equates to 180 passenger-miles per gallon – double the rail value.

Applying those fuel-consumption figures to the 2006 national rail function suggests that, if that function had been discharged by express coaches and lorries operating on uncongested motor roads, the total fuel consumption would be reduced by 24 per cent.

Estimates of carbon emissions depended on the emissions per tonne of diesel burned and on the emissions per gigawatt-hour of electricity delivered to users. Our calculations, based on the average emissions of the UK generating industry, suggest that, if the national rail function were carried out by express coaches and lorries using the railway alignments, then carbon emissions would be reduced by 13 per cent.

However, Jim Russell (formerly director general of the South Yorkshire Passenger Transport Executive) points out that any large-scale increase in the demand for electricity will increase the life of coal-fired power stations. There is discussion of the same point in the RSSB paper cited above and in the White Paper 'Delivering a Sustainable Railway'. Hence, it may be appropriate to use the emissions associated with coal-fired generation. Those emissions are double the average for the generating industry as a whole. If that approach were adopted, then, rather than reducing rail's carbon emission by 13 per cent, replacement express coaches and lorries would reduce it by 37 per cent.

These comparisons, particularly those to do with freight, have to be treated with caution because of the poor database. Nevertheless, the data do appear to show that, far from being 'green' compared with equivalent road transport, rail may be the more environmentally damaging mode of transport. Despite that, the rail industry claims the 'green' banner when, in practice, the industry either cannot or will not collect reliable data.

Excluded from the above are the emissions attributable to track provision and renewal and to vehicle manufacture. Those are considerable and tend to favour road transport.

Social equity

In the 19th century, increased use of railways gained widespread political support as a means of solving the chronic overcrowding problems suffered by the urban poor. The 'remedy' was to allow them to live further away from the squalid central areas that were close to their workplaces.

However, between 1853 and 1901, developing London's rail network displaced more than 76,000 persons, mostly the poor. Indeed, by 1851, the spread of railways was already being described in the title of a poem published in the *Builder* as an 'Attila in London'. Further, train fares were unsustainably high for low-paid

231

workers, although, under the remnants of Gladstone's 1844 Railway Regulation Act, railway companies were at least compelled to provide fares at rates of 1d (an old penny) a mile (the Parliamentary Trains). In theory, these fares would enable the working classes to commute. However, the initial effects were soon reversed and many of those who migrated to the suburbs in large numbers in the 1800s wished to return because their incomes could not sustain the higher cost of living there. (After all, one old penny in 1840 is equivalent to *circa* 30 pence at today's prices or 87 pence in terms of today's wages. Hence, in terms of wages, the cost of a ten-mile return journey was equivalent to between £15 and £20 at today's rates. Further, the 'Parliamentary Trains' were deliberately scheduled at inconvenient times and were extremely slow.)

Thus, the impact of railway expansion was to increase the size to which towns might grow, and the extent to which some social strata of urban populations could spread, but, far from providing equity between rich and poor, or relieving town-centre overcrowding, the reverse was the case.

Today, people from the top quintile of household income travel five times as far annually by train as do people from either of the bottom two quintiles. Despite the latter and the history, the railway lobby has created the illusion that rail is, in some magical way, socially inclusive. The reality is that the subsidy paid to the railways has always benefited the well-off rather than the poor. If it is the poor who are to be benefited, it would make more sense to subsidise cars than trains.*

*Except for conversions, the historical data in this section is from Leon Mannings's PhD thesis, 'From Accommodation to Constraint: an Analysis of shifts in UK Transport Policy', University of London 2004.

The costs, profits and losses

The foregoing demonstrates that, in operational terms, road outperforms rail by a very wide margin indeed, typically by a factor in the range 2–4. That would be tolerable if rail were cheaper. However, as we shall see, instead of that, rail is more expensive than road by a very wide margin indeed.

Capital for the track

During an inquiry into the West Coast Main Line Modernisation Programme, the cost rose from £5.7 billion to £13 billion. So far the cost is £8.1 billion. At that inquiry, it was said that the expenditure was to be concentrated on the core 1,000km of track. Hence, if the £8.1 billion is to be believed, the cost is *£8.1 million per kilometre of track.*

The main East Coast high-speed rail proposal was to cost £36 billion. There were to be four tracks over 796km. Hence, the unit cost was £11 million per kilometre of track. A cheaper alternative was to cost £8 billion. It would provide 322km (200 miles) of double track. That yields a unit cost of £12.5 million per kilometre of track. (Source: SRA.)

Alternatively, consider the cost of Railtrack's original modernisation programme. It ranged from £50 billion to £70 billion, excluding new high-speed line proposals. Probably most of the expenditure was to be concentrated on not more than a quarter of the track length. At any rate, the West Coast Main Line Modernisation was concentrated on only 1,000km (621 miles) of its 2,800km (1,740 miles) of track. On that basis, the cost of the national programme had the range £6.25–8.7 million per track-km.

In contrast to that, the *Independent* newspaper of 17 February

1999 reported a Treasury study that estimated the replacement cost of the M1 at £2.1 billion for all works and land, or £2.5 billion at 2007 prices. The lane length, assuming six lanes all the way from the M25 to Leeds, is 1,800 km. Hence, the cost is £1.5 million per lane-km.

The Highways Agency's estimate for a dual three-lane motorway complete with hard shoulder is in the region of £15 million per km. That includes a 45 per cent optimism bias and VAT. Stripping those out yields £1.18 million per lane-km. If hard shoulders and central reserves are included as potential lanes, the unit cost per lane falls to £0.8 billion per km. Those costs include land, 12 per cent, earth works, 20 per cent, and structures, 45 per cent.

Against that background, we have set the cost of motorway construction at £1.2 million per km of lane. That is 6.75 times less than the cost per track-km of mending and modernising the West Coast Main Line, 9 to 10.4 times less costly than the East Coast High-Speed Line, and 5.2 to 7.25 times less than the somewhat speculative estimate of the cost per track-km of Railtrack's original network-wide rail-modernisation programme.

Track maintenance

In 2006–07, the capital expenditure on strategic roads amounted to £1.4 billion, and the current-account expenditure to £1.706 billion. Previous discussions with the DfT suggest half the capital may be in the category of heavy maintenance, renewal and enhancement. Adding that to the current-account expenditure provides close to £2.4 billion. Hence, with a lane length of 50,000–55,000km, the cost per lane-km has the range £43,000–£48,000.

In contrast, rail management statements suggest that maintenance for National Rail will range between £2 billion and £3 billion per year for the decade. The track length is 32,000km. Hence, the annual cost per track-km has the range £62,500–£93,750 – one and a half to two times the cost per lane-km of maintaining the strategic road network.

Rolling stock

A typical railway carriage contains 75 seats, lasts 25–30 years and costs about £1 million. A traction engine costs several times that amount. Annual maintenance may amount to 7.5 per cent of capital. Repaying capital and interest at 3.5 per cent (the Treasury discount rate) over 30 years requires an annual payment of 5.4 per cent. Hence, the annual cost of a seat in a railway carriage (excluding the traction unit) amounts to £1,720. In contrast, a 50-seat coach may cost less than £200,000 and last at least 15 years. Repaying the capital at 3.5 per cent would cost 8.68 per cent annually. Adding maintenance at 7.5 per cent yields 16.18 per cent, providing an annual cost per seat of £650 – over 2.5 times less than the train.

Signalling

If the national rail network were converted to a system of reserved motor roads, they would operate, as do the trunk roads and motorways, with scarcely a signal in sight. In contrast, the signalling for rail costs billions of pounds. At one time a price tag of £6 billion was put on the train detection systems. That has been cut back to perhaps £4 billion at today's prices. If that is to be repaid at 3.5 per cent over 30 years and if annual maintenance is set at 5 per cent of capital, then the annual cost amounts to *circa* £400 million.

Profits from roads, losses from rail

Fares have not covered operating costs since 1955. Hence, it is unreasonable to expect loans or capital expenditure to be repaid from the fares. Instead, both capital and loans should be added to the subsidy as though both were current expenditure.

In the 20 years to 2015, the operating subsidy – capital expenditure funded by the government and loans guaranteed by the government – is likely to top £100 billion.* The £100 billion amounts to £5 billion annually. That is equivalent to taking £200 from every household in the land every year for 20 years – at a time when half of us use the train less than once a year. The subsidy is also equivalent to:

- £250,000 for every mile of railway track;
- 19 pence per passenger-mile, if freight is ignored;
- 12.8 pence per item-mile if tonne-miles and passenger-miles are added together to form one unit.

In contrast to the bottomless railway pit, the Treasury takes close to £50 billion annually from motorists. That is made up of excise duty and VAT on fuel, road tax, VAT on new cars, company car tax and Insurance Premium tax. The £50 billion is equivalent to £2,000 for every household in the land. Deducting annual expenditure of around £9 billion yields a net tax take of £41 billion, or £1,600 per household.

The £41 billion taxes (net of expenditure) may be apportioned

* Our sources are National Rail Trends, the White Paper 'Delivering a Sustainable Railway', July 2007, the proposed expenditure on Scottish railways (excluded from the White Paper), and Network Rail's annual reports.

according to the 32 per cent of all vehicle-miles driven on the strategic road network. That yields £13.1 billion to the Exchequer annually – equivalent to a tax of over £500 per household in taxes.

It is also equivalent to a 'profit' to the Exchequer of:

- £380,000 per mile of lane;
- 9.3 pence per passenger-mile, if freight is ignored;
- 6.3 pence per item-mile if (as for rail above) tonne-miles and passenger-miles are added together to form one unit.

The bar chart in Figure 16.7 compares the astonishing losses from rail with the corresponding huge profits to the Exchequer from the strategic road network.

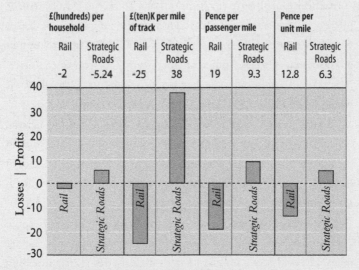

Figure 16.7: Profits and losses from roads, losses from rail

The dimensions of the national rail and
strategic road networks

The rail network is about 10,000 miles long. The track length is double that, providing some 32,180 track-kilometres (20,000 track-miles).

The width between tunnel and viaduct walls, on double-track railways, is generally 7.3 metres (24ft). That is the same as required by the carriageway for a two-way trunk road without marginal strips and verges. On straight sections, outside tunnels and viaducts, the level width of a double-track railway is 8.5 metres (28ft). On bends, the width is greater to accommodate carriage overhang. Even where there is only a single track, widths are often sufficient for double tracks.

Railway tunnels provide headroom of 4.16 metres (13.6ft) to 4.77 metres (15.65ft) above rail top. Removing the rails and some ballast would increase that by half a metre (1.6ft), providing a minimum 4.66 metres. That is sufficient for nearly all container lorries (4.3m/14.1ft) and double-decker buses (3.9–4.4m/ 12.8–14.4ft). It would be insufficient for car transporters (4.9m/16ft).

Hence, claims by the railway lobby that railways are too narrow or lack the headroom to allow conversion to motor roads should be dismissed along with demands that any such conversion should include 3-metre (9.8ft) verges, as met with on new roads built on green fields.

In contrast, the motorway and trunk-road network is some 7,500 miles long. The lane length is in the range 30,000–35,000 miles. The width of a standard lane is 3.65 metres (12ft), providing 7.3 metres (24ft) for a two-way single carriageway road. Many A-roads are only 6 metres (19.7ft) wide.

Lastly, the rail network has the advantage of serving the hearts of our towns and cities. There the railway's land is generally sufficient for a four- to six-lane motorway. In contrast, the strategic road network peters out on the fringes of urban areas.

The costs of conversion

Paving the railways would increase the lane length of the strategic road system by close to 60 per cent. A range of sources suggest a unit cost for conversion of some £100 per square metre, including all ancillary works. Applying that to an average lane width of 5 metres (16.4ft), instead of the 3.65 metres (12ft) that would be the standard, and to the track length of 32,000km yields a total cost of £16 billion for the entire system, a fraction of cost of the rail modernisation programme.

Speed and journey length

National Travel Survey data show that half of all rail journeys are less than 20 miles long and that 90 per cent are less than 80 miles long, including the journeys to and from railway stations. For all these journeys, the time taken by a car or express coach would often be less than by train, particularly after taking account of service frequencies.

Even for the truly long journey, the car dominates. Apart from cost, one of the reasons is that rail's speed advantage is eroded by the need to travel to and from rail stations and the waiting times involved. After all, theatre performances and business meetings are not scheduled to suit railway timetables.

If speed were everything, then it should be borne in mind that in the 1960s, before the imposition of speed limits on motorways, express coaches cruised at 90mph on the M1. The safety record was excellent. At that speed, the 360 miles from London to Glasgow would take four hours.

The past

Dan Pettit, chairman of the National Freight Corporation, wrote in *The Times* on 17 October 1972,

> The car and the lorry have come to the rescue of the city... the way the environmentalists in particular talk about the railways reminds me of the tale about the king's new clothes. It is an exercise in mass self-delusion ...

On 29 April 1974, when Frances Cairncross (now managing editor of the *Economist*) was the economics correspondent for *The Guardian*, she wrote,

> ... when trains are still the theme of nursery rhymes and children's stories, it is small wonder that the railways have a romantic fascination for most adults. Only years of nursery conditioning can explain the calm with which the public has accepted a bill of £3,000 millions [£35 billion at 2007 prices] to subsidise British Rail over the last decade...

The future

Let us take a peep into a future when the railway nursery rhymes and self-delusion have at last been discarded in favour of the facts. The railways will, of course, have been paved so as to provide a reserved system of motor roads.

Instead of a ten-carriage train departing periodically, specially geared coaches would leave at five-minute intervals, or when full, on busy routes. Elsewhere, the service would be tailored to meet demand with vehicles varying in size down to the minibus. Costs would be cut by a factor of at least four, death rates by a factor of two and fuel consumption by 25 per cent. Journey times would, in nearly all cases, be reduced.

In addition, the many thousands of hectares of railway land that have lain derelict in the hearts of our towns and cities these last 50 years for want of good road access would at last be developed as business parks, retail parks or industrial estates. Further, countless lorries and other vehicles would divert from the unsuitable rural roads and city streets that they currently clog, saving fuel and bringing very large environmental benefits to the people using, or living close to, the historic road network.

All of this may be had at a fraction of the cost of rail. What is required is the vision to bring this vast project into being. It should appeal particularly to those billionaires seeking lasting fame. After all, if the system were as profitable to the Exchequer as is the strategic road network, it would generate nearly £400,000 per year per lane-mile, so yielding *circa* £8 billion annually, compared with the current subsidy to rail of £5 billion.

All that is required is a puff from the right direction and the house of cards, built at such cost to the taxpayer by the railway lobby, will come crashing down. When that happens, the railways will vanish within a decade. Our great Victorian rail network will at last be brought into effective use as a system of motor roads, a project that the Victorians themselves would praise.

The truth is that the only transport policy that makes sense is this apparently controversial proposal. That is because, at least in urban areas, large-scale, low-cost road construction is impossible except on the railway alignments. Those alignments have been basking in the sun of government subsidy these last 50 years doing next to nothing for the nation. Conversion to motor roads would bring the most tremendous benefits that can be imagined, both financial and environmental.

The proposal is controversial because the nation has been fed propaganda for over 50 years. Those stories, coupled with the sentimentality surrounding the great age of (extraordinarily dirty) steam, have elevated rail to a kind of religion, virtually beyond criticism, failing only because of man's failure. In contrast, the truth is that the railways failed long ago due to the high cost and lack of flexibility inherent in vehicles mounted on wheels with steel tyres.

That failure has continued because the nation has not bothered with the arithmetic that can so easily be used to demolish the myth, and because it takes time for science to overturn belief held with the almost religious fervour enjoyed by railway enthusiasts. Hopefully, the force of the facts will make itself felt before the end of the century.

18

LIGHT RAIL, TRAMS & HIGH-SPEED TRAINS

PAUL WITHRINGTON

DOGMA
Trams provide a cheap and environmentally
sound urban transport system.

IN 1949, THE TRAMS were seen as an embarrassment to the capital's planners. In that year, Lord Latham, chairman of the London Transport Executive, delivered a speech outlining the plans for the tramways conversion programme in which he stated, 'The loss on the trams is about £1,000,000 per year' – equivalent to £25 million at today's prices.

There is, of course, nothing wrong with a tram compared with a bus except that the tram takes three times as long to stop, costs four times as much, offers little or no routing flexibility and has a fraction of the capacity, provided the bus enjoys a right of way free of congestion. Despite that, light rail and tram systems are extraordinarily popular with the planners and in the public mind.

The systems for which we have data are Docklands, Strathclyde, Manchester, Tyne and Wear, Sheffield, Centro (West Midlands) and Croydon.

- The capital cost at 2003 prices for Manchester, Tyne and Wear, Sheffield, West Midlands and Croydon totals to £1.8 billion. If Docklands light railway is added, the bill is £2.93 billion. In addition to that, Merseytram had the option of spending £225 million, Manchester hoped to spend an additional £900 million, South Hampshire was bidding for £270 million and Edinburgh for £375 million, providing a total of £1.77 billion. Adding these to the money already spent gives a total cost of £2.93 billion, or £4.7 billion in 2007 values.
- With the exception of Sheffield, no system covers operating costs. When the costs of capital and maintenance are added, the total cost is between 3.2 and 5.8 times larger than receipts.
- Including the annual cost of capital and maintenance, the annual subsidy per journey on a tram ranges from £2.50 (for Manchester) to £4 (for Tyne and Wear).
- The average one-way flow per track for the systems has a range equivalent to 92 to 525 buses per day, each bus containing an average of 20 people. That may be compared with the potential of up to 10,000 vehicles per day for a single lane of a motor road managed to avoid congestion. That suggests that tram systems make catastrophically low use of transport land precisely where that land is at its most valuable. Indeed, in highway terms, most of these systems are all but disused.
- The number of employees per car (or per tram) has the range 4.3 to 10.9.

- Average journey lengths have the range 3.2 to 10.5km (2–6.5 miles).
- The average train or tram load ranges from 17 to 50 (excluding Docklands, which has 69 passengers).
- Capital costs, including construction, per route-kilometre have the range £7.6 million to £10.4 million, excluding Docklands, for which the value is £43 million.

We have sought recent information on comparative fuel consumptions but with little success. However, data from 1990 suggested the fuel used equated to 51 passenger-miles per gallon for Tyne and Wear and 55 for Strathclyde. Data from 2003 for Croydon's Tramlink gave a figure of 92 passenger-miles per gallon (it has phyrister control – using breaking to provide energy for traction).

In comparison, buses returning 8 miles per gallon (as they may do on rights of way such as these systems enjoy) and with the average passenger loads associated with UK tram systems (the range is 17–50, excluding Docklands) would return 135 to 400 passenger-miles (217–644 passenger-kilometres) per gallon.

We conclude that none of these systems would have been built had the costs and operational limitations been properly understood. Of course, the Europeans are famed for trams. However, is it merely a case of the grass seeming greener over there? Whatever the case, Sir Terry Mulroy, at an Institution of Civil Engineers meeting held on 21 November 2002, said that '…if one asks the planners in Geneva, home of the tram, if they would do it again, they will say quietly, never again – far too expensive.' Meanwhile, '…the town of Grenoble has had to renew its tram network after only ten years.'

High-speed rail

High-speed rail is a huge area and we touch on it only superficially. The much-vaunted systems in Europe are said to be profitable. However, that is after ignoring all but so-called 'operational costs'. The capital has been provided by the taxpayer, and in France alone the total subsidy to rail is said to be around €12 billion annually.

Currently, Network Rail and others are lobbying for high-speed rail. Five such lines are proposed. Since costs in excess of £30 billion have been associated with the London-to-Edinburgh route, the five lines may, if ever built, cost in excess of £100 billion. History suggests none of that would ever be recouped from the passengers and the cost would fall on the taxpayer.

If the system were to be used as the current rail network, it would be enjoyed by the rich rather than by the poor. After all, as noted in Chapter 17, those from households in the top quintile of income currently travel five times as far by rail as do those from households in either of the bottom two quintiles.

Those canvassing for such systems do so on three grounds. First, they look over the English Channel and long for toys on the scale of the French TGV system without considering the cost. Second, they claim that such systems would bring untold wider economic benefits to the regions, overlooking the benefits that those regions would enjoy if taxes were lower or if the money were available for profitable investment instead of being poured into a certain loss maker. Third, there is the notion that high-speed rail has significantly lower emissions than competing rail systems and is far cheaper than air travel.

The global-warming scare is useful in that context, but, even if that scare turns out to be correct, (a) the presumption that high-speed rail has the lower emissions may turn out to be wrong (at high

speeds the emissions are greater) and (b) the effect of transfers
between air and rail on total emissions would, in any event, be trivial.
For example, unless large numbers transfer from air to rail, the
schedules would remain unchanged, hence the immediate effect of
individual transfers on emissions would be zero. And, as noted in
Chapter 17, large-scale increases in demand for electricity may
prolong the life of coal-fired power stations. There is an argument
that it is emissions from coal-fired generation that should be
considered in comparison to the CO_2 from aircraft. On that
assumption, the data in Table 18.1 suggest that air travel may emit less
carbon per passenger-km than would high-speed rail. In any event, if

	(A) generating industry average	(B) coal-fired generation
Pendolino West Coast at 200 kph	48.1	97.2
	55.0	110.0
Class 91 East Coast at 200 kph	41.5	83.0
Eurostar at 300 kph	68.8	137.5
Ryanair	85	
Diesel car driver only	138	
Diesel car Av occupancy	86	
Diesel car driver + 1	69	
Diesel car driver + 2	46	
Coach 30 passengers 10 miles/gal	28	

Table 18.1 Carbon dioxide − Gms per passenger-km

the emission argument is to be taken seriously, rather than either of those, we should either stay at home or go by express coach.*

Furthermore, the data do not take account of the very substantial carbon emission that should be associated with construction. For example, the carbon dioxide emission associated with 1km (0.6 mile) of double-track rail (providing four rails) is 240 tonnes for the crude steel and perhaps 480 tonnes after adding for subsequent manufacture. The distance from London to Edinburgh is 632 km (392 miles). Hence, for the rails alone we have an emission of *circa* 300,000 tonnes of carbon dioxide if the facility were double-track, or 600,000 tonnes if four-track. To that must be added the effect of the fastenings, the sleepers, the overhead electrification and all the other engineering works. Hence, the immediate effect would be a substantial increase in emissions that would take years to recoup even supposing rail does lead to a lower year-on-year output.

Against that background, we trust the nation will not impose more tax pain upon itself by pouring £100 billion into a glorious toy such as high-speed rail.

---

*(a) The fuel consumptions of the trains in the above were from Professor Roger Kemp. (b) The RSSB report 'Traction Energy Metrics' suggest a 60 per cent load factor for coaches, yielding 30 passengers in a 50-seat coach. (c) A coach would achieve 10 miles/gal on a motorway alignment.

4

QUESTIONABLE DOGMA

19

RECYCLING RUBBISH

EAMONN BUTLER

DOGMA
We can save the planet by recycling our rubbish.

A FOUR-YEAR STUDY by the UK's Environment Agency concluded what many of us, with our back-of-the-envelope calculations, had long suspected: that disposable nappies are no worse for the environment than washable ones.

It seems obvious that washables are best. You just wash and reuse them, instead of throwing them into a plastic bag and letting them pile up in landfill sites, leaching nasty stuff into the soil. But, when the Agency looked at the whole life cycle of washables and disposables, it found 'little or no difference' in their environmental impact.

Amazing? Not at all. Washing nappies uses water – that precious resource – which you need gas or electricity to boil. You use bleach and detergent, which goes down the drain and into our waterways. Disposable nappies are bulky to transport, but the cotton in the washable ones we use in Britain is brought into the country over long distances from China, Pakistan or America. If you send your nappies to the laundry, you have to think about the pollution, noise and traffic congestion caused by the laundry van. Add it all up, and disposables are no worse for the environment and if you burn them as fuel they are much better.

When you've finished with a polystyrene cup and are about to pitch it into the bin, you probably feel faintly sinful, and embarrassed that you didn't ask for a washable ceramic mug instead. Well, don't be. It takes a lot more energy to make that ceramic mug, and then each washing uses yet more energy, plus water and detergent. In fact, Canadian chemist Martin Hocking calculated that you would have to use the mug 1,000 times before the energy consumed per use fell below that of the foam cup. And, if you drop and break it before then, it's no contest. And then, because they are clean, you are much less likely to pick up nasty bacteria from a foam cup.

Don't get me wrong. I am an avid recycler. I recycled the steel spoon and the china bowl I had my breakfast from this morning. Tonight, I'll be recycling my clothes for use another day. I do this because the market sends me a clear signal – these things are expensive to replace, so reusing them makes sense.

But my local authority wants to go further. It insists I have four different plastic bins into which I separate things such as glass, plastic, newspapers and vegetable waste. I would not mind if I thought it did any good. But most of these things are quite worthless. I'm ordered to wash out my glass jars, which of course uses up water, electricity and my precious time. Glass is made of cheap and plentiful sand: transporting it to recycling centres adds to the amount of diesel fuel we burn and pollution in the air. I cannot really see that sand is a limited, precious resource. The same for my garden sweepings, which, quite literally, grow on trees.

Some years ago, Franklin Associates, an environmental consultancy that undertakes work for the US environmental and energy authorities as well as for private companies, pointed out that extensive recycling like this was, on average, 35 per cent more expensive than conventional rubbish collection, though it could be more than double. It found that Los Angeles required twice as many vehicles to manage its recycling policy, adding of course to the city's already dismal airborne pollution.

In another bin I have to put my old newspapers. But reusing old newspapers is not without its own cost either. People forget that recycling is itself an industrial process. Newspapers are covered in ink, so turning them into useful paper and packaging means using various noxious cleaning agents, not to mention power and water. A US government study found that recycled paper actually contained more toxic substances than fresh paper, and that the pollution caused was roughly the same. Much of the recycled paper ends up being burned.

In fact, the attempts to make us give up plastic and adopt paper may be entirely misconceived. Studies suggest there is no obvious environmental advantage to using a paper cup rather than a polystyrene one, for example. It depends on whether you're more concerned about the pollution from the petroleum used to make polystyrene or the soup of water and chemicals that is used to make the paper. It's hard to balance these different environmental effects in any objective way.

Nevertheless, the UK government is trying to browbeat supermarkets into ditching their plastic carrier bags and is encouraging them to use paper ones instead. But plastic bags use 40 per cent less energy to produce. Paper manufacture produces 50 times the water pollution. Plastic bags don't use a significant amount of oil to make (far, far more oil is simply burned to produce heat and electricity or to power vehicles). And it takes 90 per cent less energy to recycle a tonne of plastic than a tonne of paper.

If you don't recycle but just throw your bags into landfill, you'll find that paper bags take up five times the space of plastic ones. Our newspapers may have been getting a lot fatter, but our shopping bags have been getting a lot thinner – today they are just a quarter of the thickness they were when we started using them in the mid-1970s.

The same is true of other packaging – and for good commercial reasons. Aluminium, for example, is expensive. It takes a vast amount of energy to process bauxite into the metal, and to mine it in the first place. It's why aluminium is one of the few things that definitely *are* worth recycling, not that my local authority makes much of a fuss about it. An executive from Coca-Cola once told me that his company saved a million dollars for every micron it shaved off the thickness of a can. The result? Well, 30 years ago, crushing a drink can in the hand was seen as a feat of great strength. Now a small child can do it.

Old batteries contain lots of scarce metals that are expensive to mine and purify and toxic when put in landfill or incinerated. They are ideal for recycling, yet few councils encourage it or even provide the facilities for it.

Though we are always being rebuked for the amount of packaging we throw into our bins, modern packaging is in fact very efficient, both economically and environmentally. The average household in the United States produces a third less rubbish than the average household in Mexico, largely because Mexico's bins are full of food waste and America's aren't. Where meat, for example, is centrally processed and packaged, almost the whole of the animal is used: the parts that we don't care to eat are processed for pet food or produce oil and fat for other purposes. Packaging the meat from 1,000 chickens requires around 7.7kg (17 pounds) of packaging, but saves around 907kg (2,000 pounds) of waste products going into our household rubbish. The same with vegetables, orange juice and much more – the skins and peelings can be used, instead of being transported into our homes, discarded and transported out again. And, when food is properly packaged, consumers not only save time, but are also less likely to contract food poisoning. So let's not be ashamed of our throwaway plastic packaging.

The reason why local authorities are so determined to get us to recycle is not because it makes sense. It is that European Union legislation forces them to do so. The EU Landfill Directive aims to cut the waste ending up in landfill sites. It aims to reduce the amount of biodegradable waste in landfill to just 65 per cent of its 1995 level, and every European country has to comply.

A lot of people argue that we must recycle because we are running out of holes to put our rubbish in. That is hardly true. In 2001, the US Environmental Protection Agency calculated that an area of 26 square kilometres (10 square miles) would comfortably hold all the rubbish that America will produce over the 21st century – which in a country of around 9 million square kilometres (3.5 million square miles) does not amount to much. Finding holes isn't really the problem.

True, there are other concerns with landfill sites. In the past, noxious substances have leached from them into neighbouring land and watercourses. However, the worst examples stem from the illegal dumping of industrial waste – a problem of human frailty rather than a problem of the existence of landfill sites. Leaching from modern landfill sites is minimised by constructing them in dense clay and lining the walls with plastic membrane with heat-sealed joins, and keeping water out by covering all parts of the site, except those immediately open for business. And, while landfill sites do indeed produce methane, a 'greenhouse gas', the fact that sites are kept dry and the fact that we are throwing away less rotting food and more lightweight packaging all helps to curb it.

A lot of the compulsion to 'reduce, reuse and recycle' comes from people looking only at what goes into their bins rather than what is produced throughout the whole lifetime of a product, including its manufacture, transportation and use. That is why we blithely assume that reusable nappies are environmentally superior to disposable ones, ignoring the environmental cost of making and washing them. That is why we feel guilty when we throw away packaging, forgetting that packaging allows us to use food and other resources much more efficiently. And we often forget the environmental cost of the recycling process itself – something that itself produces various kinds of pollution.

We are also burdened with outdated or mistaken ideas about what is and is not biodegradable. Early supermarket bags were tough and strong, and certainly did hang around for years. I still have papers tied up in plastic carriers from the 1970s. And people still believe that plastic bags last for ever. But, in fact, modern supermarket bags fall to shreds in just a few weeks. They are no good for wrapping and storing stuff at all. On the other hand, we all assume that paper just rots away, as it does when you leave some out in the rain. But, in the dry, airless conditions of a landfill site, it can last for – well, centuries, probably.

We also assume that almost everything is worth recycling, whereas in fact most rubbish is simply rubbish. Recycling looks cheap because the things that are worth recycling are already recycled, for profit. Tens of millions of tonnes of scrap metal, for example, are reclaimed each year by industrial companies, simply because it makes commercial sense. When I lived in America, students went round collecting old beer cans because they could earn a few cents apiece just on the scrap value. But you cannot extrapolate from that and say that everything is worth recycling. The value of most of our rubbish is so low that it just does not make sense to bother.

The same is true of our notion that recycling must be good because it saves valuable resources. But, on those grounds, it would make no sense to recycle paper when the temperate forests of the northern hemisphere are growing faster than we harvest them for paper. For most of the last 20 years, commodities of all kinds have been getting cheaper rather than more valuable, so the same applies there. And, where things do get scarce, market prices give us a clear signal to cut back our use of them, or seek out alternatives. Modern cars, for example, are lighter and use far less fuel than those of 20 years ago. Meanwhile, a rising oil price makes it more economical to exploit the reserves that are located in more difficult places, and to invest in new forms of energy.

The market, likewise, should tell us when it is worth recycling things. Is it really worth all the energy, transport – and pollution – costs involved? Or the human effort? Years ago, when labour was cheap, goods were built to be repaired. Now, when they go wrong, we just chuck them away. We sacrifice the physical resources to save another important resource, one much overlooked in this debate, our own time.

People have recycled things for millennia. Recycling is practical and essential. It can save money, time and resources that we can use for more valuable ends – but only when it is informed and voluntary. Policies that force us to recycle actually end up squandering things that we should sensibly simply discard.

20

UNSUSTAINABLE ARGUMENTS

AUSTIN WILLIAMS

DOGMA
Our present way of living is unsustainable.

IN 1979, *Time* magazine ran an article entitled 'The Deluge of Disastermania', quoting historian Christopher Lasch's observation that 'storm warnings, portents, hints of catastrophe haunt our times ... Impending disaster has become an everyday concern.' He was right to challenge such miserablist crystal-ball gazing, which he described as a pathological 'sense of ending'. Unfortunately, we have sunk much further than even Lasch could have imagined.

First, it is worth noting that random predictions of imminent collapse or disaster are nothing new. A cursory glance at some of the key moments in the recent history of catastrophism provides a flavour of paranoia throughout our age.

Paul Ehrlich's book *The Population Bomb*, written in the 1960s, was a strident demand for population restraint in order to avert ecological disaster. He prophesied a global die-off by the mid-1980s that would bring the world's population down to an acceptable level of around 1.5 billion people. By 1972, the Club of Rome's wildly inaccurate but seminal 'Limits to Growth' predictions foretold a looming energy and resource crisis (which, according to environmental commentator Bjørn Lomborg and Oliver Rubin, 'had a profound impact, spawning alarmist headlines'), and proposed that we work within the 'physical constraints of the planet'. And lately, Jared Diamond's bestseller *Collapse* brings a range of populist concerns together when it speaks of 'deforestation and habitat destruction, soil problems (erosion, salinization, and soil fertility losses), water management problems, overhunting, overfishing, effects of introduced species on native species, human population growth, and increased per capita impact of people'. However inaccurate the models used to achieve these rhetorical conclusions, the 'environmental calamity' *zeitgeist* seems to have been around, in various guises, for a lot longer than we might imagine.

Compare Diamond's litany of human-induced environmental damage with Anton Chekhov's *Uncle Vanya*, first published in 1897:

Man has been endowed with reason, with the power to create, so that he can add to what he's been given. But up to now he hasn't been a creator, only a destroyer. Forests keep disappearing, rivers dry up, wildlife's become extinct, the climate's ruined and the land grows poorer and uglier every day.

This oft-quoted passage has such a contemporary resonance that it is often cited as a forerunner of modern environmentalism. So, is the contemporary context of 'unsustainability' just another version of the old story? Well, if we unpick the 'context' it becomes clear that, while the *'form'* of environmental concern remains the same, there are some clear and worrying differences between now and then. For example, Chekhov's character, Astrov, is reflecting on the destructive state of imperial Russia, where, even in that parlous period of history, 'what sustains him is faith in the future'. Indeed, the seldom-quoted follow-on line has Astrov saying, 'If roads and railways had been built in place of ravaged woodlands, if we had factories, workshops and schools, the peasants would have become healthier, better off and more intelligent.' Unsurprisingly, he wants development.

The Chekhov lines show that societies – however repressive – and their populations have historically viewed the future with anticipation and ambition. Unfortunately, that is no longer the case. Modern-day environmental commentators rail against infrastructure and certainly wouldn't be seen to endorse roads and factories (the Ford plant at Dagenham – a road-based factory – is described by one environmental journalist as helping to create a 'foul atmospheric soup'). Today, the built environment is decried, and unsullied nature lauded. Admittedly, in the past there was criticism; today there is merely exasperated despair.

In the new millennium, social criticism, such as it is, is born of a *fear* of change rather than being an integral part of the *process* of change. As such, for some environmentalists, while they talk of the 'future', their fears have become rooted squarely in the present. They don't really want to take society backwards but, at the same time, they are petrified of the future. Whether it's George Monbiot or Gordon Brown, Al Gore or Gore Vidal, there is an acceptance – and a promulgation – of the idea that the future is a worrying place to be. For the first time in history, stasis is the preferred political option and 'future-proofing' is the new buzzword.

Indeed, a variety of concerned environmental lobbyists have suggested that we could have only 'a few years' before the plug begins to be pulled on humanity. The ever-reliable alarmist Andrew Simms of the New Economics Foundation suggests that there is a 'ticking clock of 100 months', while some newspapers suggest that it is already too late. Dr James Hansen, the man credited as the first leading scientist to announce that global warming was happening, says that we are on 'the precipice of a planetary tipping point... if we go over the edge... there will be no return within any foreseeable future generation'. The assumption – and the dogmatic assertion in many instances – is that geological timeframes breed complacency and undermine the blameworthy human scale of the problem. And so, when it comes to current debates about global warming these days, reference is seldom made to Kyoto's 100-year timescale, implying that 'a century' lacks the desired rhetorical immediacy for action.

But what is this action that we are meant to take to avert disaster? Well, in almost every discussion, it seems that the answer is 'restraint' of some form or another. To a certain extent, the required action is inaction! British environmentalist George Monbiot summed it up when he hoped that his book would 'make people so depressed about the state of the planet that they stay in bed all day, thereby reducing their consumption of fossil fuels'. Whether it is carbon rationing, limiting resources, minimising impacts, conserving energy, or simply reducing, reusing and recycling, the essential message is that we must do less, and *be seen* to be doing less *with* less. While this might sound like sensible efficiency measures, it actually cuts to the heart of the belief in our ability to change things.

Not so long ago, humanity was rightly seen as a source of creativity, progress and dynamism. In those (dim, distant) days, many people recognised that things were not as good as they could be, and aimed to transform the world for the better, striving to overcome natural constraints to human and social development. Since then, social and political critiques have gone off the agenda. In the last ten years or so, society has begun to believe in the immutability of nature and seeks to blame so-called wasteful and irresponsible lifestyles as the cause of all manner of global ills. This, by its very nature, is corrosive of the very notion of human-centred solutions. Chris Rose, ex-director of Greenpeace, chides us about 'our near-suicidal devastation of the environment – from filling our atmosphere with greenhouse gases to seasoning our food chain with toxic chemicals'. Jared Diamond speaks to the broader contemporary social malaise, implying that mere human intervention in nature is dangerously insipient. In his words, 'ecocide has now come to overshadow nuclear war and emerging diseases as the threat to global civilisations'. It all has a terrible contemptuous fatefulness to it.

It is now common to hear hopeless environmentalists wallow in their belief that 'we are the first generation to knowingly hand . . . our planet on to our children in a worse condition than we received it'. Not only is it not true (or meaningful), but it actually patronises parents, all of whom still want a better world for their children. But, with misanthropic criticisms ringing in their ears, a parent's instinct for a better world is often matched by their lack of belief in the possibility of its realisation. The dangers of such sentiments are even more profound than that: these days, aspiring to a better world is frequently deemed to be materialist hubris and, hence, inherently problematic. The noted economist Jeffrey Sachs of Columbia University's Earth Institute finished his Reith Lecture at Peking University in Beijing by suggesting that humans had reached a point of 'literal unsustainability'. It is now so acceptable to argue that humanity is a problem rather than a solution, and that nature is pre-eminent, that it is not uncommon to hear the assertion, quoted here from *New Scientist*, that 'humans are completely living beyond their ecological means'.

For those readers not old enough to remember, it wasn't so long ago that we weren't governed by ecology. Nature was something that was to be tamed, where possible. Admittedly, nature was to be enjoyed as well, but ultimately we were the species with dominion. So, for instance, it used to be blindingly obvious that the underdeveloped world needed development and that its harsh, unforgiving environment was something to be overcome. At that time, anyone with an ounce of decency condemned the lack of development in the Third World and recognised that a solution to those struggling in Third World conditions was to improve their lot – their material wealth and their living standards. No longer.

So influential has the people-as-problem mindset become in the West that, compared with our misanthropic age, the Reverend Thomas Malthus almost comes across as a nascent liberal. Sociologist Frank Furedi spells it out well when he writes,

Malthus' reservations about the human potential were a product of his deep-seated hostility to the optimistic humanism of his intellectual opponents: Condorcet, Godwin and others. And yet, he made it clear that despite his pessimistic view of population growth 'it is hoped that the general result of [my] inquiry is not such as to make us give up the improvement of human society in despair'.

Unfortunately, despair almost appears to be the political ideology of choice these days, as society sinks further into a slough of despond and environmentalists find that it has become increasingly difficult to argue for a positive future. The moralistic hectoring beloved of environmentalists is beginning to pall and, increasingly, environmental advocates are relying on the partial use of (impartial) science to give them legitimacy and to pressurise us into action. In this way, 'sustainability' is being turned into an equation, albeit a political equation. Just as the phrase 'the science shows' has been used by climate-change activists to close down debate (as we can see in this book), so environmentalists are buying into 'sustainability' as a simple, unequivocal, mathematical numbers game that will force us to agree.

Trying to give social sciences an immutable scientific gloss is a way of hiding a political agenda behind the safety of 'data', and providing 'unequivocal justification' for whatever policy is on offer. For example, Porritt legitimises his authoritarian one-child policy with mathematical clarity: it is, he says, 'the cold, unyielding calculus of climate change' that shows why there should be fewer people. The 'scientific' logic is straightforward: if humans cause carbon emissions and carbon emissions are a problem, then the fewer people there are breathing out, using energy and motorised transport, then the fewer emissions there would be. This Vulcan logic might suit environmentalism, but we should not be taken in by it.

If humanity is really reducible to a crude discussion about its carbon emissions, then sustainability advocates have a shallow view of fellow human beings. Forget culture, art, creativity, inventiveness, emotion or intelligence. Environmental science is now being fashioned into a policy agenda that can tell us what to do. Paul Ehrlich is even trying to go further. A generation after his 1968 book *The Population Bomb*, he is now trying to get 'culture to change its course' and is looking for data that can 'fit it into a model that will tell us what direction to move in'. Quite explicitly, he hopes that evidential history will provide a route map for the future.

There are thousands of nonsensical 'research' papers out there, but one typical 'scientific' report sets the scene. Apparently, according to the UN initiative the Millennium Ecosystem Assessment (2007), 'shrimp on the dinner plates of Europeans may well have started life in a South Asian pond built in place of mangrove swamps – weakening a natural barrier to the sea and making coastal communities more vulnerable'. The obvious environmental conclusion is that we should preserve the mangroves by eating fewer shrimp. But actually, even if the scientific evidence did show a relationship between eating food and rising sea levels, it doesn't necessarily make it legitimate, or imperative, to eat fewer shrimp.

Another argument would be for more shrimp to be consumed, especially if selling more shrimp will raise the local economy to such a degree that profits could be channelled back into conservation and protection projects – or, even more importantly, into human-centred development schemes. And then again, perhaps, the taste of shrimp is well worth the alleged damage to mangrove areas. Perhaps local people can't stand swampland and are glad about its demise. Perhaps natural sea defences that are reliant on mangroves are inadequate for a modern world, anyway.

On the last point, history has shown that the poorer a

'community' is, the more vulnerable it is to natural disasters. It is worth remembering that, on a regular basis, floods kill thousands of fishermen and families living in shoddy dwellings on the Bangladeshi waterfront, whereas people living in the Netherlands are reasonably unaffected by rising tides. Both countries are at, or below, sea level but one has sufficient infrastructure to protect it (set in train after the devastating floods of 1953), while the other has nothing. The survival rate reflects the level of real development. So, while evidence of rising tides can give us useful information, it cannot − or ought not to − govern action. The answer to one of Bangladesh's problems is massive infrastructure spend, rather than the introduction of carbon-traded solar panels in the vain hope that it will hold back the flood tides.

There are many answers to sociopolitical questions that mathematical parameters do not address. But, without challenging this way of thinking, reports by unelected environmental spokespersons will continue to conclude in the call for restraint: 'If we want to change the world, we must force governments to force us to change our behaviour.' NASA scientist Dr James Hansen correctly notes that 'public and policy makers must consider all factors in making decisions and setting policy. But these other factors should not influence the science itself or the presentation of science to the public.'

However, he seems unable to take his own advice and is perfectly at ease with scientists making policy pronouncements: demanding a moratorium on coal-fired power stations, transforming vehicle manufacturing, carbon rationing and the obligatory call for more funding for climate research.

The truth is that politics should stay out of science, and vice versa. Unfortunately, Rajendra Pachauri, who chairs the UN Intergovernmental Panel on Climate Change (IPCC), told the *Observer* that, in order to counter climate change, people should 'give up meat for one day [per week] initially, and decrease it from there'. He has also added a suggestion that we 'ride a bike and be a frugal shopper'. Science writer Fred Pearce agrees, and suggests that one solution to climate change will be 'the emergence of an older, more mature, less frenetic, less consumerist, and more frugal society'. Surely, society could develop more ambitious solutions to global problems than this kind of bunker mentality. This might appease the ambitions of the survivalist fringe, but this is no recipe for a humane future.

Indeed, in a rhetorical (and possibly real) environmental 'war economy', it is hardly surprising that there are casualties in the drive towards frugality. Achim Steiner, executive director of the UN Environment Programme, says that 'the human population is now so large that the amount of resources needed to sustain it exceeds what is available at current consumption patterns'. This straightforward Malthusian position argues that resources cannot grow sufficiently to cater for the ever-increasing population – a self-fulfilling prophecy given the rebuttal by environmentalists of erstwhile efficiency-raising genetically modified (GM) crops – but Steiner's condemnation centres on unsustainable 'consumption patterns'. This normally refers to the developing world's aspiration for 'Western' or 'improved' consumption patterns, i.e. their hopes for more than low-calorie 'sustainable' subsistence consumption inherent in peasant economies.

Boris Johnson, just before becoming the Mayor of London, said that the 'primary challenge facing our species is the reproduction of our species itself'. He went on to describe the sight of densely populated Mexico, seen from above, as like 'a horrifying vision of habitations multiplying and replicating like bacilli in a Petri dish'. Johnson has often been chastised for his race gaffes; however, describing Mexicans as bacteria didn't seem to warrant a rebuke, given that such a contemptuous view of humanity is fairly established among even the most vociferous politically correct campaigner. Professor John Gray, for instance, thinks that 'human life has no more meaning than the life of a slime mould'.

Once it is conceded that humanity is 'the' problem, all manner of misanthropes can crawl out of the woodwork. *Guardian* journalist Madeleine Bunting says that there are simply 'too many people', while James Lovelock is fêted for suggesting that the world is suffering a 'plague of people'. Reith lecturer Jeffrey Sachs is applauded for speaking contemptuously of the world 'bursting at the seams', and Chris Rapley, head of the Science Museum, says that he is 'not advocating genocide [, just] ways to reduce the birthrate'.

Environmental grandee Jonathon Porritt spells out the new dogmatic certainties: 'It's blindingly obvious,' he says, 'that completely unsustainable population growth in most of Africa will keep it permanently, hopelessly stuck in deepest, darkest poverty.' The missionary zeal with which Establishment environmental spokespersons want to eradicate the numbers of poor people – in order to *help* poor people – is troubling, as is the unquestioning climate in which such Victorian population control edicts have been able to gain acceptability. It is hardly surprising that the Optimum Population Trust is rearing its ugly head. A recent article concluded that 'the real concern for all of us is not that the populations of Africa, China and India are increasing but that as they grow in numbers and wealth they are developing a Western model of living: increasing consumption and escalating expectations'.

For those, like me, who believe that new life is a wellspring of potential creativity rather than damage, of hope rather than harm, the mere concept of too many people is an alien one. The re-emergence of collective self-doubt and of population restraint should be worrying for those who believe in human potential and human exceptionalism.

Actually, the essential problem here is that such a self-flagellating criticism of the benefits of development and of progress itself leads to self-doubt, confusion and an inability to deal with big questions – those that even a concerned environmental sensibility poses. Unsurprisingly, much of this paralysing uncertainty finds expression in the paradoxical excitement and paranoia around the inexorable rise of China and India. On the environmental side, it is now common to hear explicit fears about the Chinese building hundreds of coal-fired power stations to fuel their economic growth; or scares about India's revolution in cheap flights and cheap cars. It is ironic that the dynamism of these emerging economies is reflected – within their own country – in their faith in the creative, rather than the destructive force of humanity.

In fact, it is 'growth' and 'development' (without being prefixed by 'sustainable') that are the very things that are needed in order to lift poor populations out of penury. It is progress and technology that are required to raise people's capacity to live beyond subsistence, not subsistence that determines the number of people that can adequately be supported. The more that we accept the dogmatic orthodoxy that says that we are in thrall to nature, and the more we eschew material growth, the less humanity will be able to shape the future. Under the mantra of 'protecting nature', phrases such as 'human impact' are nowadays seen to be inherently negative; words such as 'development' are unconsciously conjoined with adjectives such as 'harmful', 'detrimental', 'destructive', 'dangerous' and 'unsustainable'. Maybe we shouldn't be surprised when the *Ecologist* reports that 'there are increasing calls from within the environmental movement for greater use of force, rather than less, including pro-active armed intervention on a military scale . . . in order to safeguard the planet and its resources'.

While Britain sees population growth as a potential source of instability, China sees its increasing population as a source of productive capacity. Britain is in the throes of cultural pessimism, China is riding high on optimism. While China has been reported to have relaxed its nefarious one-child policy (long condemned as part of the West's ideological assault on Chinese totalitarianism), Jonathon Porritt has demanded a one-child policy and 'zero-net immigration' in the West.

At the very least, it is mean-spirited to portray humanity in such a contemptuous manner; and uncharitable to suggest that the fewer of us that exist, the more there is to go round. But, as this chapter has tried to show, at its heart is a suggestion that consumer patterns, materialism and modern standards of living are the problem. Such is the acceptance of this reactionary position that even commentators like the editor-in-chief of the erstwhile Open Democracy discussion forum says, 'Malthusian realities will lead us away from energy-intensive lifestyles and consumption-intensive values.'

Similarly, the Worldwide Fund for Nature (WWF) doesn't want to 'distract attention from the fundamental problems inherent to consumerism' and to 'question the dominance of today's individualistic and materialistic values'. Whether it is celebrity chef Jamie Oliver bemoaning the fact that cheap food exists, blithely ignorant about the needs of those less well off than himself, or Elizabeth Farrelly, who condemns our 'unchecked appetites', in essence, their dogmatic appeal for restraint is considerably worse than attitudes of the brutal neo-Malthusians. At least they want to level *up*. The new breed of frugal environmentalists advocate actually want to level *down*.

The point is this: even if humans are responsible for carbon emissions, and even if carbon emissions are responsible for global warming, and if global warming is a major threat to society, it still doesn't follow that humans should use less carbon-intensive energy sources. As an example, the rapid improvement of the Chinese and Indian productive economies – using carbon-intensive means – has lifted vast numbers out of poverty, brought them into the productive sphere and potentially allowed these people to participate in the intellectual practice of problem solving.

It is about time we challenged the moralistic parameters of sustainability, rejected the cod-scientific notion that 'the debate is over', challenged the piffle that masquerades as social policy justification, deplored the idea that humanity is a problem – or that growth and development are harmful – and started arguing for a bigger picture. Once those parameters are set, we'll find that 2 billion heads are better than one.

21

THE ETHICS INDUSTRY

STANLEY FELDMAN

> Ethics: relating to morals; a set of moral principles,
> a code of moral conduct.
> —Oxford Dictionary

DOGMA
We need codes of ethics to distinguish right from wrong.

SOME YEARS AGO, I watched a group of desperate women and teenagers, in Guatemala City, scavenging in a pile of rubbish for something saleable or usable. They had been employed at a nearby 'sweatshop' making shirts and tops for a Korean company. Protests from the 'ethical trading' establishment had forced the government to clamp down on the exploitation of these poorly paid workers and this had led to the closure of the factory. Without any money coming in, the women were reduced to scavenging and begging.

A similar story can be told about the effect of these so-called 'ethical' pressures on the welfare of indigenous populations from many poor countries.

In the name of ethical standards, the poor have been reduced to begging.

In the past 50 years or so, there has been an explosive growth of what can be loosely described as 'the ethics industry'. Ethical committees, ethical standards, ethical foreign policy, ethical employment, ethical travel, ethically sound products, etc. have mushroomed. Today, ethicists seek to control what we do, how we shop and how we think. Large companies, banks and trading organisations are obliged to have an officer in charge of 'ethical standards'. These self-perpetuating, self-appointed guardians of our moral compass have projected their own, often very personal, beliefs on to the public without proper scrutiny of either their effects or their values. Unfortunately, the consequences of the well-intentioned efforts of these bodies are very often the opposite of the 'general good' they seek to promote.

The problem arises because of the inevitable assumption that what is right or moral in a particular circumstance should be so in every case. Because what is right and what is wrong often depend upon the circumstances, they cannot, and should not, be codified. There are no strictly acceptable rules as to what is moral and what is sensible that are universally applicable at all times. No matter how eminent the authority, any attempt to make such a code is doomed to failure. The rigidity implicit in all so-called codes is likely to cause perverse decisions. It often does more harm than good.

Possibly the earliest code of ethics – the rule book we are told we should all obey – is the Ten Commandments. This is deemed to have the highest authority: the authority of the God of the Old Testament. Even God's ethical code does not stand up to scrutiny.

Thou shalt not kill. But is there no such thing as a just war?

Even God commanded the Israelites to kill the Amalekites, to destroy Jericho and turned Lot's wife into stone. Was this really ethical?

If you are attacked, should you not defend yourself? If we accepted the 'shalt not kill' dictum as part of a code of ethics, no person and no country would be safe from those who did not subscribe to the same code.

Thou shalt not bear false witness. Should telling a lie be made an offence? Certainly Plato thought lies were justified if they were for the public good. Bertrand Russell faced this dilemma and concluded there was nothing immoral or wrong about telling a lie if it resulted in doing good. He cited the example of being confronted by a man carrying a dagger demanding, 'Which way did she go?' Not to 'bear false witness' could lead to a death. It certainly should not be a moral obligation never to tell a 'white lie'.

Thou shalt not covet they neighbour's ox? Is it really unethical to aspire to better oneself, to seek a better education for your family or to try to emulate your neighbour's higher standard of living when yours is inadequate? Isn't competition to be lauded? Should you not seek the provision of water, sanitation and warmth if it is available to your neighbour? It is hard to believe that it is ethically sound to say to the poor in India that they should stay poor and not seek a way out of their miserable life in a shanty town.

If even God can't get it right, what hope is there that mortals sitting in judgement on important-sounding ethical committees can do so?

In his Reith Lectures, many years ago, Professor Ian Kennedy, a guru of ethics, took issue with the right of any doctor to make a life-or-death decision for his patients. Such a decision would be based on the doctor's knowledge of the patient's particular disease and the wish of the patient and his relatives. Such a course of action, he told us, was unethical and illegal. Kennedy's cry was that no doctor should play God. But someone has to make these decisions. What he was actually saying was that it should be lawyers, like himself, who should play God in these situations.

Undoubtedly, there are occasions when the legal route is the right path to travel, but to make this an ethical commitment, a codified, legal obligation, rather than an individual decision based on what is sensible and humane in a particular circumstance is wrong. It fails to recognise the imperative that any decision should take account of any mitigating factors. It should not only be in a patient's best interest but it should take account of his or her wishes. It is possible that the wishes of the family and the urgency with which many of these decisions have to be made in a particular circumstance may influence a compassionate decision. One cannot ignore the degree of suffering involved and the cost incurred in taking the legal route. Every situation is unique. To subject all patients in an urgent or painful situation to a legal inquisition before a medically sound and socially desirable decision is made, in the name of ethics, is immoral.

Let us take the example of a premature baby of less than 1kg (just over 2 pounds) bodyweight. Babies this small usually require artificial ventilation and intravenous feeding to keep them alive until their lungs mature sufficiently and their gut starts absorbing nutriment. Even then, a large proportion of them die and many of the survivors will have pulmonary complications. The moral question of whether or not to embark on the long expensive resuscitation with the hope of producing a viable child who may be severely handicapped cannot be settled by hard and fast rules, ticking the boxes of ethical treatment. To embark upon artificial ventilation and feeding could clearly be looked upon as striving officiously to keep the baby alive and interfering with nature in a perverse rather than benevolent way. If the mother is an immature 15-year-old child who had caused the premature birth by trying to end the pregnancy, or had attempted to commit suicide, the attitude to resuscitation of the baby should surely be different to that of a similar baby born to an older mother who had spent years trying to conceive but who has had multiple miscarriages. It is clearly wrong to impose some sort of 'ethical' straitjacket on the decision-making process. A doctor's moral duty is always to do whatever is in the best interest of his patient – in this case, the patient is the mother.

Ethical products

There are ethical products that claim they are produced entirely from natural sources. From toiletries and soaps to shirts and furniture. Indeed, various chains of high-street shops claim this is sufficient reason for expecting you to pay extra for their merchandise. But why is it unethical or immoral to use chemicals from a bottle to produce a toiletry, or cotton grown conventionally in order to produce similar products often chemically indistinguishable from the so-called ethical ones? We accept that the machines that make these products are made of metals and chemicals. No one would suggest using an 'ethical' wooden apparatus to produce the so-called ethical goods. It is a marketing device to increase the profits of the manufacturer.

It is wrong to suggest that one particular form of merchandising somehow gives a product a superior or moral value because it is labelled, by the producer, as 'ethical'.

Most people feel a sense of unease about the rearing of chickens in battery conditions. Of course, it would be nice to have them roaming freely, but is it unethical to use this method of producing chicken protein at a price that is affordable, even for those on very low incomes? In a world in which the cost of first-class protein is likely to become increasingly expensive, is it morally right to demand that the only chicken meat on sale should be from birds reared in such a way that their meat becomes too costly for the poor to be able to afford? A dilemma of this nature cannot be solved by ticking boxes in an exercise in ethical standards: it has to be solved on a case-by-case basis according to the economic imperatives.

Is it morally permissible to displace a small local population or threaten the habitat of an endangered species in order to build a dam or a facility, such as the Severn Barrage, that will provide lighting and heating to many thousands of homes?

These questions are not those of right and wrong; they are not ethical issues. They are political matters involving analysis of the overall economic imperatives and the social and environmental costs before final decisions are made.

Sitting safely on the back of an elephant, watching tiger cubs playing with their mother, is a fantastic experience. The thought that tigers are an endangered species and may one day become extinct makes one angry. However, if I were a subsistence farmer, living on the fringes of a national park, I would not hesitate to shoot a tiger that destroyed my precious crops and attacked my children. Would I be acting unethically in protecting my children?

Unfortunately, today, decisions about endangered species are viewed through the myopic vision of the specialists of the ethics industry, who see the immediate problem as a moral one in which the risk to an endangered animal takes precedence. Many of the animals on the endangered-species list reflect their emotional appeal rather than the actual risk of their extinction. Would anybody be seriously concerned if the smallpox bacillus or the *Anopheles* mosquito became extinct? Would they seriously wish that dinosaurs still roamed our countryside or that sabre-toothed tigers lived on the South Downs?

While it is reasonable to protect endangered species and to avoid disrupting people's lives unnecessarily, each decision should be made on its merits. Preserving the animal kingdom's status quo or frustrating change should not be an 'ethical' issue.

Should all workers performing a similar task be paid the same? At first sight, this seems reasonable, but it fails to take into account the very different costs of living. Those advocating an international level playing field in employment are pushing for this to become an ethical issue. This has spawned the 'fair trade' industry and the concept of ethical investment. Provided labour is freely offered and freely purchased, competition will invariably cause the rewards of labour to be paid at the going rate. This will inevitably be higher in some countries than in others according to the cost of living. Even in the UK it is recognised that, due to the higher cost of living in London, there is the need for 'London weighting' of pay.

Foreign policy

Is it possible to have an 'ethical foreign policy?' If so, what does it mean? Should we have soldiers who do not kill? Do we make treaties only with those who share our views of the world? Is it moral to deny military assistance to those being abused if they do not agree with our code of ethics?

Who decides what is an ethical cause and what is unethical? Should we invade Muslim countries to make sure they treat women and homosexuals in a way we consider 'ethical'? Our government protested to China at the unethical treatment of the Dalai Lama but failed to do anything to stop millions being killed in the Cultural Revolution. Is it ethical to bring to justice those Serbian commanders who we believe have committed crimes against the Bosnian population when we do not indict Robert Mugabe for the terror and death he has brought to the people of Zimbabwe?

Surely the first objective of the foreign policy of any government is to further the interests of its countrymen. This may, at times, mean supporting governments whose moral code we do not share. After all, in World War II, the USSR was our ally against the evils of Hitler, although it was a repressive, murderous regime. The concept of an ethical foreign policy is insupportable.

Slavery is uncivilised, it is 'unethical'. Is it moral to prevent children being sold into slavery in countries such as Sudan if the alternative for them is freedom and starvation? It is circumstances, not statutes, that determine what is right and what is wrong.

The 18th-century philosopher David Hume pointed out that it is inevitable that our attitude as to what is morally right depends on how close our relationship is to those involved. There are few mothers who would not see a man who kidnapped their child subjected to torture if it were necessary in order to see their offspring returned unharmed – and few would blame them. If it is right in the case of a mother and her own child, why not for any child, even that of someone in a remote and foreign land?

Duress and torture

Is it immoral to subject anyone to 'mental or physical duress' in order to obtain information that can save thousands of lives? We see it as a cause for satisfaction when the leader of a terrorist gang is killed. Why then is it so much worse if, instead of killing him, he is subjected to duress and allowed to live? I wonder, if one could ask the terrorist if he would rather be deprived of sleep or killed, which he would prefer.

Cruelty is certainly against all ethical codes. If one accepts all forms of physical or mental duress to be unacceptable – a sort of Gold Standard of ethics in international law – then it becomes legal and ethical to sacrifice thousands of lives to protect one person from this pressure, even if it saves him from being killed. Surely this is wrong. Torture that includes acts of cruelty, sanctioned for political reasons, for revenge or personal benefit is obnoxious but subjecting a known terrorist to physical or mental duress for humanitarian ends surely should not be universally condemned. All civilised people abhor torture but to codify it as 'never ethically acceptable' is to contradict one's moral duty to save life.

Ethics in hospitals

Nowadays, all hospitals have 'ethics committees'. It is sensible to prevent the rights or the safety of patients being needlessly endangered. Having been involved in research long enough to see it develop from being the sole responsibility of the person organising the research to the corporate responsibility of the hospital, I can only applaud the change. The trouble is that there is no yardstick against which the ethical standard can be set. As a result, too often ethics committees take an absolutist approach, assuming that all research is unethical until proved otherwise.

This negative attitude results in unnecessary expense and delays. It is not helped by the difficulty of interpreting various statutes such as that on confidentiality of information and the requirement for patient permission for retention of blood and tissue specimens. In this way, a simple, anonymous comparison of a patient's ECGs becomes an ethical problem involving confidentiality. The collection of blood, urine or other bodily fluids for diagnostic purposes is a clinical, not an ethical, decision, but their retention for research requires the informed consent of the patient.

It becomes impossible to conduct retrospective studies, as the patient's consent would not have been obtained at the time of the fluids' collection. There is little doubt that many of the diagnostic procedures and treatments practised today would never have seen the light of day if they had had to run the gauntlet of an ethics committee. Because of the rigidity of the ethics rule book, research is impeded and, it can be argued, future lives are put at risk.

The retention of tissues from autopsies for later study into the nature of disease was standard practice before the Alder Hey Hospital pathology 'scandal', and without it modern medicine would still be in the Dark Ages. Human anatomy would still be based on the anatomy of the pig and histopathology as we know it today would never have developed. The actions of Dick van Velzen, the pathologist involved in retaining the internal organs of dead infants for later study into the cause of their death, may have been excessive or even imprudent, but surely it was not immoral. After all, our sense of morality evolves, like most other things, and what would once have been quite acceptable becomes less so in the fullness of time.

The nationalisation of morality

The ethics industry has nationalised morality. An individual can no longer make a free choice as to what is right and what is wrong. It is a matter of ticking boxes. If sufficient ticks are in the right boxes, it becomes ethical. It is algorithmic morality. It destroys the fundamental differences between man as an individual and man as an animal in a herd. He can no longer choose according to his own conscience but must conform, at all times, to the diktat of authority.

If ethics is synonymous with personal morality, then the way it is implemented by the ethics industry has lost sight of this fact. The problem is that what is moral is a personal choice between right and wrong in a particular circumstance. Once one tries to introduce a legal codification, the element of personal decision making, agonising over the pros and cons of each case on its particular merits, is lost. It is time we saw these 'ethical bodies' for what they are, a way of absolving oneself from the duty of having to make difficult, and at times distasteful, choices. It is an abrogation of personal responsibility.

The growth of the ethics industry has resulted from a tilting of the balance of decision making from the individual to the state or an authorised organ of the state, under the influence of pressure groups of dubious probity. It has downgraded personal responsibility and individual decision making to an act formulated by a committee. There is little doubt that Dr Mengele's experiments would have gained the approval of an ethics committee consisting of Nazi sympathisers. It is time to stand up to the unethical tyranny of the ethics industry and make people responsible for their own actions.

22

GLOBALISATION

STANLEY FELDMAN

DOGMA
Globalisation impoverishes the poor and
makes the wealthy richer.

A historical perspective?

WE THINK OF globalisation as a 20th-century phenomenon that
has put power into the hands of supranational businessmen and
reduced the democratic control of political decisions. To the left-
wing politicians, this is an unholy alliance bringing together
financial self-interest with the levers of political power. To the right,
it is the logical extension of the free market – an economic
necessity. The two views are not mutually exclusive but they
prompt the question: is it a good thing for the person in the street?
To answer this question, one needs to look at the historical
development of supranational power.

One can make a case for globalisation to have been started by St
Paul when he journeyed through Asia Minor spreading the gospel.

Was this not a sales pitch for what had been a local religious undertaking, an attempt to sell a particular brand of belief to a multinational audience? If Paul was the initiator of this, then St Peter was its first CEO. Thanks to the Emperor Constantine, the global headquarters for what was to become a multimillion-dollar empire was established in Rome, following the takeover bid at Nicene. Over the years, it has spawned various brand names that have established the hegemony of this worldwide empire. It has browbeaten rulers, ruthlessly suppressed opposition and influenced the decisions of democratically elected governments. It runs by its own set of rules and declares those who leave its control to be apostates.

Today, the green movement has assumed the mantle of a global company. In many ways, it has replaced religious beliefs with the green catechism. Like the church, it has given rise to many splinter organisations, each interpreting this catechism in its own peculiar way. Like the church, it defies detailed analysis of its beliefs and labels disbelievers as heretics. It has enormous influence and it proselytises constantly. It claims to be the only true faith and one that has universal approval. It ruthlessly suppresses the opinions of those who disagree with its dogma; it labels those who oppose its views as 'deniers' or in the pay of sinister multinationals; it positively seeks to influence democratically elected governments and to make supranational decisions over the heads of sovereign states. It is truly a global empire of enormous power.

Beside these mega-supranational sharks, other global enterprises appear as benign minnows. The principal difference between them and commercial global industries is that their motivation is conviction rather than profit (although many fortunes have been made on the coattails of these enterprises), reward in another world, not in the present one. Yet, for all their avowed good intentions, there can be little doubt that both of the mega-supranationals have set out to bend the decisions of democratic institutions and governments to their own ends. They have pursued campaigns that have ultimately caused death and misery to thousands and they have denied the advance of scientific reason.

The Inquisition, the Crusades and the persecution of alternative faith sects in the name of religious belief hangs as a heavy burden on the faith-led multinationals, while the pressure on governments from the green movement caused them to ban DDT after Rachael Carson's book *The Silent Spring*, although this was based on supposition, biased reporting and unsubstantiated beliefs. The result was millions of preventable deaths from vector-borne diseases such as malaria.

They have opposed nuclear energy in a world desperately short of its benefits; they campaign against genetically modified foods, even though hunger is widespread among the poor. The inability of religion to heed the scientific evidence of Copernicus and Galileo of a heliocentric universe has parallels today in the refusal of the green movement to countenance many technological advances. Like a religion, they advance their own improbable theories and persuade their followers that death and disaster will follow unless they follow their particular creed or eat only food that they certify as 'organic'.

Both of these organisations insist that their view of the world is the only one that is true, and they deny and malign the doubters. They spread their influence by word of mouth, instilling fear of the future, of sterile oceans, famine and disease, the flooding of London and New York by the rising seas or by the teaching of damnation in the world to come. They proffer redemption and salvation to those who follow their credo. Unlike other global companies, they provide only comfort and reassurance as the reward for buying into their products. How strange, then, that these globalised industries, selling their improbable beliefs that call for sacrifice and discomfort, do not attract the same opprobrium as others, which offer valuable products and services that make life easier and more pleasant but trade for profit, when both have, on occasions, demonstrably caused harm on a massive scale!

Few would argue that commercial multinationals have completely clean hands. Instances of exploitation of the naive are common among multinationals – from the African slave traders to those today who are exploiting the mineral wealth of Africa and the cheap labour of Asia. Such exploitation was also common among the missionary zealots of the church in their conquest of South America and Africa and in the green movement's attempts to deny Africa the benefits of pesticides, modern farming and electricity in the name of sustainability.

An indifference by multinationals to the impoverishment caused by switching production from one area to another with a cheaper wage structure or of employing labour in the developing world at minimal rates of pay is undoubtedly a cause for concern. But, then, the church was never slow to extort money from poor communities to build monumental branch offices, or cathedrals; nor were greens in their advocacy of expensive organic food, without any evidence of its superiority, and their refusal to countenance genetic modification of seeds that would make crops more productive, cheaper and richer in essential vitamins. Truthfulness and transparency have often been lacking in the performance of commercial multinationals, but, then, few people have been accorded access to the secrets of the Vatican; and the spokespeople for the Green movement (like St Paul) frequently proclaim that their message is more important than truth. While one should not ignore or forget the 'dark' side of all supranational enterprises, ultimately they must be judged on their present record. Are they a force for good? Have they improved the lot of ordinary people?

The good

Like the Church and green supranational bodies, the implicit objectives of globalisation are exemplary. The economy of size, coupled with free access to worldwide markets, makes it possible to develop and deliver goods and services to all at an affordable cost. It allows research and development of new technologies, medicines and knowledge that could not be afforded by a local company, however wealthy. It stands for the common good, the improvement of the lot of the common person, and a better and fairer distribution of the world's riches.

Like the Church and the green movement, it has undoubtedly made progress towards these aims in many regions even if it has failed miserably in others. Like those of the Church and the green movement, its efforts have from time to time been suborned by human error, misdirection and personal ambition. It is not the principle of globalisation that is at fault: it is the result of those responsible for its application in a particular area who fail to see it applied in a sensible manner. Those who seize on examples of the failings of globalisation to undermine the credibility of the whole movement confuse the 'baby with the bathwater'.

Most Indians today reject the Gandhi philosophy of 'for every home a loom' – the economics of the cottage industry. It is one that would for ever tie the peasant to low-paid employment. The burgeoning of the Indian middle class has been the result of education to meet the demands of a market-led economy. As a result, their skills can be used to compete successfully in the world, aided by a low, but rapidly rising, cost base.

A visit to Bangalore, always a fairly prosperous Indian town, is to see the global economy working. Today, it is an international centre with excellent colleges, hospitals and housing. Visit parts of Utter Pradesh, a region that has turned its back on market forces and international investment, and one finds major roads that have degenerated into tracks covered with potholes, poor housing, few hospitals, irregular electricity supply and few educational facilities. It remains a largely feudal society without any expectation of better things to come.

In the China of Mao, with his fierce antipathy to outside investment and his doctrine of every village with its own iron foundry, millions died of starvation. An industrious nation was strangled. While not everyone has benefited from the investment that has replaced the old regime, today China is developing into a global power, and education, welfare and prosperity are spreading. Those who hanker for the romantic idyll of the 'good old days' and would have us return to a pre-industrial civilisation do not appreciate the squalor, degradation, poverty and disease with which it was associated.

Only 50 years ago, the expectation of life in northern Europe was around 45 years. Today it is approaching 80 years. Globalisation has brought modern medical treatments and drugs, good sanitation, access to electricity, clean water and better nutrition, all of which have contributed to a longer, healthier life. In many developed countries up to half of the children go on to further education; many now study in foreign universities. The telephone is taken as a right and a communication system that 50 years ago was exclusively for the rich is now enjoyed by all, thanks to global investment.

In Victorian times, foreign travel was restricted to the wealthy who could enjoy the educational benefits of seeing the world. Today, mass travel, a global enterprise, has made it available to huge numbers who take advantage of it to holiday in parts of the world visited only by wealthy adventurers and explorers a hundred years ago.

Television and radio bring the wonders of the world into almost every home. It can be enjoyed by the Inuits in the Arctic Circle and in the shanty towns of Mumbai. Globalisation has made a variety of foods and goods that were previously found only on the tables of the very rich, available to all.

The biggest impact of a global economy has not been on the rich. They could always buy the services, the travel and the goods they wanted. It has been of greatest benefit to the poor, to the less well-off and to those of middle income whose lifestyle today lacks few of the amenities that were previously the provenance of only the very wealthy.

The opening up of previously closed home markets to international companies has produced a movement of money and employment to many countries who were in need. It has provided funds to build the infrastructure of modern cities. It has produced global markets of sufficient size to produce the economies of scale and to allow the huge investments necessary to make complex, expensive technological developments possible. It results in competition for hitherto closed markets, driving down the cost of finished products. It has become an article of economic faith that this is a movement for the good of all. Why, then, has it caused resentment and protest?

The bad

It was the World Trade Organisation meeting in Seattle in 1999 that brought the anti-globalisation protestors on to the streets. At the time, the mass movement of short-term money had so disturbed the capital markets in 1998–99 that a global depression was imminent. It united a community of the losers in a common assault on what they saw as a club of the rich who were exploiting those less fortunate who lacked the financial and industrial muscle for meaningful competition. They felt impotent.

The protestors included poor farmers put out of work by imports from subsidised factory farms of wealthy countries; owners of small businesses that could not compete with the purchasing power of the multinationals; local industries locked out of reasonably priced loans by banks that preferred the security of the global enterprises; workers whose jobs have been exported to areas where labour is cheaper; those concerned that drugs that could save lives were too highly priced to make them available in vast areas of the world; and a loose alliance of the romantics who hankered for a simple life – the days of the 'loom in every home'. To some of the protestors, it was the hegemony of the dollar, the power of the US, that brought them out in protest. There was a general, ill-thought-out resentment at the idea that everything, including native cultures, was up for sale to the highest bidder.

There is no doubt that the political implications of the global economy had led to mistakes being made. Too often, exploitation resulted from one of the partners to a trade agreement holding all the cards and levers of power while the other operated in a blinkered atmosphere of hopeful expectation. Such unbalanced competition invariably results in disappointment, a failure of proper competition, and in some instances unfair exploitation. Instead of the benefits of globalisation being evenly spread, they became restricted to those already rich. The benefits failed to trickle down and too often it meant that the poor became poorer.

This was seen particularly in Africa, where it often resulted in the exploitation of the poorest in society. Even if the poor did not actually lose, there was resentment that they had been excluded from the benefits of globalisation. These ill feelings frequently led to social instability. A particular problem in 1999 was the uncertainties caused by the rapid withdrawal of capital from the developing markets of SE Asia, where there was a sudden surge in unemployment. The failure to insist on sustainability and the lack of control of the flow of capital revealed a political fault line, rather than a failing of globalisation. Even more recently, the failure of the banking industry to control the excesses of the market in debt has led to an atmosphere of fear and suspicion between institutions that has frustrated the normal ready access to capital necessary for people to buy a home and for the needs of industry.

It is easy to see these examples as an inherent fault of globalisation, particularly if you are the victim of the system. Most of the shortcomings are not inherent in globalisation but in the politics of world trade and the rules by which it is played in each country. It is largely a failure of the bodies established to prevent the misuse of globalisation. Much of the blame for the present financial problems in the developed world lies in the way in which individual governments have failed to supervise adequately the indebtedness of their banks. Many of the problems in the developing world lie with failure of international bodies, such as the International Monetary Fund (IMF), the World Bank and the World Trade Organisation (WTO) to understand the different priorities of developing countries. These organisations have worked on economic models that have failed to make allowance for the various forces that interact in complex and diverse systems, especially when they are put under pressure by world events. They too often assume that the economic climate will be the same tomorrow as it was yesterday and that the economic priorities of Africa are the same as those of China and South America. We can now see that the rules used by the IMF and WTO were unduly proscriptive and the lack of flexibility that was endemic resulted in a one-pattern-fits-all approach that did not meet the different conditions in individual countries.

The problem is different for the developed countries, where there has been an unduly relaxed attitude to the supervision of the complex derivatives being traded by the banks. These regulatory shortcomings pose difficult problems in our ever-changing times, but they are largely surmountable, although it seems that personal greed and ambition will always find some way to circumvent the rules and produce a degree of inequality.

The balance

It is unthinkable that the world will willingly give up the enormous benefits of global trade. It has improved the lot of the less well-off in many countries, has brought hope and prosperity to SE Asia and India, and has the potential to eliminate the scourge of extreme deprivation, provided there is the political will and the foresight to implement it in a way that will lead to a more equitable world. It is essential that the inequality of bargaining power between the rich and poor countries should not be allowed to lead to unilateral exploitation. The world today is a richer, better-educated and healthier place, and, thanks to globalisation, these benefits are available to the many rather than just the wealthy few who could afford them. The ultimate aim of globalisation must be to make them universally available.

Against this background, the pleas of those who protest at globalisation and of those who advocate a massive reduction in industrialisation in case it causes the ice caps to melt are trying to turn back the clock of progress. In such a world, the rich countries will be able to purchase absolution while the poorer countries will be largely excluded from economic advancement. This has the potential danger of bringing the supranational green movement into conflict with the demands of the developing world. Without globalisation and increasing world prosperity, people's expectations of a better, healthier and richer life will be frustrated, causing political unrest and social instability.

23

EFFICIENCY OR EFFECTIVENESS?

VINCENT MARKS

DOGMA

An efficient organisation is an effective one.

HAVE YOU EVER WONDERED why, when you go to a major train station to buy a ticket, you always have to stand in a queue? Or why, when you call a public utility, government, local authority or any other large organisation or business by telephone, you always have to hang on, having been told, 'Your call is important to us'? Or why you have to play the guessing game as to which number on your telephone keyboard you should press to be connected to a person who can deal with your problem?

It is all down to that Holy Grail of business: a quest for efficiency. This is a term borrowed from engineering, meaning 'production with minimum waste or effort', but is currently a buzz word meant to convey the impression that an organisation is effective. A machine is efficient when it produces its desired effect with the smallest amount of waste; an organisation is efficient when its employees and its capital are fully occupied for most or all of the time. It's a far cry from meaning that it's effective. Whether it is effective or not depends on the effect it has on its customers.

The replacement of individual queues at checkout points by a single consolidated one has been adopted by many large organisations and is an attempt to improve both efficiency and effectiveness. In some situations, it works – but not at train stations, where there is invariably a queue for tickets. The fact that there is nearly always a queue points to the indifference of the railway company to its customers' needs.

The longer a customer waits, the more disgruntled he becomes. He can do nothing about it – at least for the present – but another time he may decide that the car, plane or bus is quicker. What he usually can't do is withdraw his custom there and then – that would be perceived to be cutting off your nose to spite your face.

'Your call is important to us'

Something similar to queuing at checkouts occurs when making telephone calls to national and local government departments. The telecommunications industry, public utilities and large organisations, including hospitals, have persuaded themselves that, by installing automated telephone-answering systems, they have made themselves efficient. The user of the service, on the other hand, who may find it impossible to match his requirements with the list of options, and finds it frustrating, time-wasting and too often unable to provide the information needed.

'Efficiency' is a much-abused word. Too often it is confused with 'effectiveness'. Effectiveness is defined as producing a desired effect or intended result. Customers are not concerned with an organisation's efficiency in the strict meaning of the term, only with its effectiveness, even though efficiency savings are likely to be reflected in the cost. Just because an organisation is efficient, it does not make it effective, nor does it necessarily save money.

Let us take the example of the queue at the railway booking office. If, throughout the day, there is an average of four people queuing for a ticket, it means three are wasting their time. In effect, it is equivalent to a loss to those three people, who could be working productively and earning money. The real cost of the rail company's efficiency is three people's salary. In effect, the railway's meagre efficiency saving has not been transferred to the consumer, and the cost has been multiplied threefold.

Ensuring that there is always a queue at the ticket office means that a number of customers are left waiting for anything up to ten or even twenty minutes before they receive attention. This might persuade them to use the automated ticket machines that have multiplied in stations over the years but are still so comparatively crude that they are able to deal with only 90–95 per cent of customers' needs, leaving a significant number who still need personal help and have to join a queue. So long as there is a significant queue, efficiency is being bought at the expense of effectiveness.

Inconvenience and cost to travellers – now called 'customers', presumably to enhance their status – is the price they have to pay for the company's efficiency. Since in the long term all the travellers have to be dealt with and sold tickets, the net effect is that a lot of people have wasted a lot of time. The assumption that the greater efficiency shown by the company is reflected in a lower cost to the customer, as far as I am aware, has not been demonstrated.

Take, as an example, the telecommunications industry. It is becoming almost impossible to communicate with them, except by telephone. When you do attempt to telephone, you will have to go through the obligatory reminder that, to make your task easier, you will be given a choice. This is meant to ensure that you will be put through to the person best able to help you as quickly as possible. Sometimes, if the information can be expressed numerically or by voice recognition, it is possible to complete the whole transaction without speaking to a living person, and this is fine and a proper use of automation. If it can't, you will, after spending several minutes going through a question-and-answer monologue masquerading as a dialogue, eventually end up listening to a recorded message thanking you for the call and reminding you that it is important to them. Once again, a gain, in terms of efficiency by the industry, has resulted in a loss to the consumer.

Often you will be told that, due to unprecedented demand, all of their assistants are busy and you should either call back later (at some very poorly defined time) or continue to hold because – as you have already been told – 'Your call is important to us.'

How long you have to wait until you begin your genuine dialogue varies enormously in both time and cost. You are almost invariably charged from the time your call was answered by the robot until it ends, either abruptly when your frustration at hanging on gets the better of you or you speak to someone whose stated aim in life is to please you.

Existing frustrated customers of a telecommunications business – for example, those whose service has broken down and whose appointment for an engineer's visit is unceremoniously broken at the last minute – may find it extremely difficult to gain satisfaction.

A personal case study

The following is a typical example of the sort of problem that could have been resolved with one simple telephone call if only the appropriate structure had been in place for real efficiency.

The saga began with an essential piece of equipment failing to perform. Without it, I could access neither my television nor my broadband connection. My telephone link remained intact. I managed to book an appointment a week ahead for a technician to replace the faulty piece of equipment. I made a three-hour round trip to keep the appointment. Thirty minutes or so before he was due to arrive, the technician phoned to say there was no point in his coming, as he did not have the replacement part, that he had not had one for several weeks and there was no prospect of his getting one in the future. I thanked him and asked him to get a senior manager to telephone me to explain what was going on and advise me what to do.

When, several hours later, I had heard nothing from the company, I entered the parallel universe of automated answering machines. After becoming increasingly frustrated, I eventually spoke to a real person. I asked whether the information I had been given by the technician was correct and if so why I had not been told this when the appointment was made.

The assistant I spoke to was quite unable to deal with my query. She was equally unable (company policy) to put me through to anyone who could. She promised that a manager would contact me before the end of the day. No one did, so I wrote to the CEO and sent an email to say that I was seeking another supplier, as I considered they had broken their contract. I received a telephone call in response to the email next day and confirmed that the contract was void from the time they had ceased to provide service.

The company's internal communications were so poor that I continued to receive invoices for services they were unable to supply for a further three months. They ceased only after there had been an exchange of letters, two within 24 hours, from different departments of the same organisation.

Such examples are all too common and can largely be put down to different departments in a single organisation striving for efficiency without any regard to the final outcome or effectiveness.

The strategy adopted in the name of efficiency by the two industries used as examples is not confined to them. It is rife throughout modern business, which has sold its soul to algorithmic management based on computer programs designed to cope with the commonest and uncomplicated problems it encounters, but is totally unable to deal with anything that falls outside the system's parameters.

The ability to use tools to increase the range of things we are able to do is one of the things that distinguish mankind from many other living creatures. The ability to make and use machines that carry out a series of sequential actions is, however, unique to the human species. It has reached its apogee with the computer, upon which we have all come to rely. Its ability to fulfil a task is only as good as its program and the end user. These two factors impose limitations on the ability of computers to deal with problems that were not allowed for in their programming.

Machines and computers

Machines have made modern life possible. They also constitute a threat to the livelihood of those who rely upon their (limited) ability to perform tasks that require a minimum of intellectual input. Computers, like all machines, have reduced the need for human beings to carry out repetitive (routine) tasks with well-defined beginnings and ends. They almost always perform better (more efficiently) than human beings, as they do not become tired or bored, but they flounder when confronted with the unusual. This will invariably happen in real life, from time to time, and requires human intervention.

In the case of railway-ticket offices, machines can dispense (routine) tickets as efficiently and effectively as a person sitting in a booth in most instances. This has led to the closure of the ticket office for most of the day in many small stations, where the business is tidal. There are, however, occasions when ticket machines can't replace live operatives.

At major railway stations, the answer is simple: keep open sufficient ticket offices to obviate the need for long queues. This will tax the boss's managerial skills and his ability to decide which is the more important: short-term (cost) efficiency or long-term customer satisfaction (effectiveness). At smaller stations, the answer may lie in having ticket collectors on trains who can exercise discretion rather than apply a diktat that they have to follow slavishly.

Exception reporting

No algorithmic program – no matter how well written or meticulously operated – can cover every contingency. Some method of exception reporting should therefore be written into all programs. They should have a default connection. Among telecommunications and other organisations, including government departments, doing all or most of their business by telephone or email, a few do have the facility to allow the caller to get through to a cognisant individual. They can often resolve the caller's query or problems, especially if they are intelligent, willing and enabled by their superiors to do so.

This is a truly efficient and effective use of staff time. Unfortunately, with few exceptions, the most aggressive (efficient) companies often make it impossible for their skilful salespeople to connect you to other parts of the companies' internal networks.

Organisations that either do not have exception reporting or do not enable their employees to take decisions based on circumstances (evidence) may think they are efficient but they are far from effective. They risk losing customers to more user-friendly suppliers and cause needless frustration and resentment.

Every organisation will, from time to time, disappoint some of its customers. How well it deals with them will determine whether they remain customers or look for other suppliers. Although it is much more cost-effective to keep an existing customer than obtain a new one, many organisations alienate their existing customers by prioritising efficiency over effectiveness and by not ensuring that complaints are dealt with expeditiously. The popularity of television programmes such as *Watchdog* and of newspaper and magazine consumer-complaints columns demonstrates how rapidly seemingly insoluble problems encountered by individuals become resolved once the searchlight of adverse publicity is focused on them. This indicates how little attention organisations give to exception reporting.

One way of overcoming this perennial cause for complaint and the loss of actual as opposed to potential customers would be to reinstitute the practice of having managers readily available to sort out any problem.

Unfortunately, all too often the necessity to provide a quality service is sacrificed on the altar of efficiency. Too often, efficiency savings, especially those that involve a reduction in the number of managerial staff, merely transfer the cost of sorting out any problem to the customer. Organisations should seek to resolve problems before the problems become complaints, and should seek efficiency savings that do not put an additional burden upon the consumer. It saves dissatisfied customers from having to go to consumer watchdogs for help and prevents an inordinate amount of angst for the customer.

24

POPULATION

VINCENT MARKS

DOGMA
There is no need for population control.

HUMAN BEINGS, as we understand them today, originated somewhere in eastern Africa around 70,000 years ago. They spread throughout the world over the next 50,000 years. It is believed that their numbers increased very slowly until just three or four hundred years ago. They then started to increase rapidly, although this occurred to a greater extent in some parts of the world than in others. What kept the number of people so comparatively low for so long is far from certain. A limited supply of food was undoubtedly a major factor, while early death, from disease and predators, is also believed to have affected the rate of growth of the population.

For most of his early history, man was a hunter–gatherer and his impact on the environment was in no way different from that of other animal species. This all changed with the shift to agrarianism.

Groups of farmers formed settlements, communities and villages. This eventually led to the Industrial Revolution.

Agrarianism began about 10,000 years ago. In Europe, the need for land to grow food, either for direct consumption or, more wastefully, for fodder for animals that provided meat and dairy produce, led to destruction of the forests that had previously covered much of the land. It was accompanied by the elimination of many animal species. Something similar happened all over the world at various times and is still happening in those parts of the world we describe as 'developing'. When it does, it is met by protests from those who are intent on preserving an unchanging environment, as though only the one they are living in were 'natural'.

Energy and food

The ability to produce food to order, instead of relying on chance, made it possible for communities to settle, enlarge and produce larger communities, towns and eventually cities. These in turn spawned new industries, such as building, transport and heating, all of which require a constant supply of energy.

Until a couple of hundred years ago, energy was obtained almost exclusively from biomass. Wood was used for heating and the manufacture of buildings. Grass and grain were fed to the horses that were used to provide transport. Wind and water power made a small but significant contribution to the energy requirements. It was not until coal mining put a seemingly unlimited supply of usable energy at man's disposal that an industrial revolution became possible. The fact that this Industrial Revolution happened during an Age of Enlightenment, when the grip of the Church on the way that people thought and behaved was broken, is probably not a coincidence; nor is the fact that it coincided with a leap in the number of human beings surviving to adulthood.

The mining of metals such as copper, tin, iron, gold and silver began in antiquity. It was not until coal made the manufacture of iron and steel possible on a large scale that it ceased to be a cottage industry. The impact of these developments on the environment was enormous but occasioned little comment until after World War II.

The number of people

Counting the number of people living in a country is as old as civilisation itself. Censuses were conducted in Biblical times, or earlier, for military and taxation purposes. They are now almost universal.

Thomas Malthus, a cleric and economist, was a highly respected Englishman and fellow of the Royal Society – the leading scientific body in the country. In 1798, he wrote one of the most important books of all time, whose main theme is as valid today as it was in 1798. In contrast to the optimism of religious bodies and authority, Malthus's primary concern was that, as the number of people surviving to adulthood increased, they would outgrow their food supplies and eventually breed themselves into extinction. His gloomy forecast has not come true, largely because he could not foresee the amazing increase in food productivity that has occurred as a result of the 'green revolution'.

Although Malthus stressed that 'the ultimate check to population [growth] appears to be availability of food', he pointed out that this is never 'the immediate check, except in the case of actual famine'. The immediate checks he had in mind were the deaths of infants as a result of poor housing and hygiene, and infection and the deliberate limitation of family size by delay in marrying and sexual abstinence. Contraception as we understand it today was not an option.

So powerful was Malthus's message that, just a year or so before the green revolution, resulting from the introduction of scientific methods into farming, Paul R Ehrlich wrote *The Population Bomb*, a book that first appeared in 1968. In it he wrote, 'The battle to feed all of humanity is over. In the 1970s and 1980s hundreds of millions of people will starve to death in spite of any crash programs embarked upon now.' He went on to say, 'I have yet to meet anyone familiar with the situation who thinks India will be self-sufficient in food by 1971'; and 'India couldn't possibly feed two hundred million more people by 1980.'

Ehrlich has been proven wrong, although this may turn out to be only a matter of timing. The improvements in plant technology, imitated by the Nobel Laureate Norman Borlaug, enabled food production to keep up with the growth in world population. However, some of this effect has been offset by humanitarian efforts to prevent internecine wars and deaths from starvation and disease. It remains to be seen whether new agricultural techniques continue to keep up productivity to match population growth.

Today, the world population is around six times larger than in Malthus's time. Some 20 per cent live in the developed world and enjoy a high standard of living, Of the remaining 80 per cent, a few live in luxury but most live in poverty, with more than 50 per cent at little more than subsistence level.

The green revolution may well have reached its apogee, but it is not necessarily over. The invention of genetically modified crops capable of higher yields and of growing in previously unusable environments holds hope for the future. The irrational opposition to them from selfish, short-sighted people, mostly living in Western Europe, is misguided. The crisis of food production and competition between the use of land for biofuel and food in 2008 has made governments in Europe re-examine their inflexible policy towards genetically modified organisms that have a potential that is still a long way from being fulfilled.

The world's population increased slowly for the first 50,000 years. It had reached only around 250,000 by the beginning of the modern era (zero CE). Since then its growth has been rapid. Whereas the world population was probably around a billion or so in Malthus's time, it is now, only two hundred years later, some 6.5 billion and forecast to reach 9 billion by 2050. The effect of this increase on the availability of minerals and fossil fuels and on the pollution of land, sea and atmosphere was highlighted by Ehrlich as well as by D H Meadows in *The Limits to Growth* (1972), which, though hopelessly wrong on timing, was surely right in principle. Writing in 1990, Ehrlich said 'Humanity in the 1990s will be confronted by more and more intransient environmental problems: global problems dwarfing those that worried us in the late 1940s. Perhaps the most serious is that of global warming, a problem caused in large part by population growth and over-population.' Again he was prescient: right in principle but wrong on timing.

We have seen a period in which man's ingenuity and scientific knowledge has allowed him to prevent the catastrophe envisaged in the book but it remains to be seen whether there is a limit to man's resourcefulness in meeting the needs of the huge, rapid increase in population in the future.

The forecasters of doom in the 1960s were wrong, because, like so many before them, they overlooked man's ingenuity. A classic example from former times was the forecast that London would be feet deep in horse manure by mid-20th century if the growth in vehicular traffic increased at the same rate as it had done during the late 19th century. No one foresaw the invention of the internal-combustion engine, which made horse-drawn buses and carriages obsolete. Nowadays, we are not so concerned about horse manure as about CO_2 emissions, the exhaustion of oil and mineral deposits and environmental pollution.

Forecasts of the increase in population have mostly been fulfilled. It is too early to say whether the slowing in the rate of population growth, forecast to occur in the next 50 years as fewer children die and the standard of living rises leading to a fall in the birth rate, will happen. If it does, it is projected that the world population will stabilise at around 9 billion or may even shrink.

The limits to population growth

Why do populations grow and what are the consequences? Malthus, who was not without critics in his own time or subsequently, wrote,

> It is an obvious truth, which has been taken notice of by many writers, that population must always be kept down to the level of the means of subsistence; but no writer that the Author recollects has inquired particularly into the means by which this level is effected: and it is a view of these means which forms, to his mind, the strongest obstacle in the way to any very great future improvement of society.

The question that should be foremost in our minds is: how can we reduce the birth rate to match the death rate? This would produce a more sustainable world population. The reasons the problem is not being faced are complex and political.

Population size

The number of people alive in a geographic locality is, at any one time, largely determined by the number of babies born who survive to adulthood compared with the number of adults who die during the same time period. Changes due to immigration and emigration are generally small in comparison. There are important historical exceptions to this, such as the immigration by Europeans into North America during the 17th and 18th centuries and emigration of people from Ireland in the 19th century.

In the 18th century, those people given unlimited land and resources doubled their numbers in just 25 years. These rates of growth were seen in America and Canada and parts of Europe. Improvements in sanitation and the provision of wholesome food and potable water, and improved obstetric care, combined with various vaccination and immunisation programmes, led to a marked reduction in infant mortality from infection and malnutrition in the developed world. This resulted in a growth in population. It was gradually counterbalanced by measures developed to reduce the number of babies conceived. By the second half of the 20th century, many developed countries had achieved negative population growth. In developed countries, this was a cause of concern that there would be insufficient people of productive age – arbitrarily set at between 20 and 60 years – to support the population and their children and the ever-expanding number of elderly people. (Elderly is arbitrarily defined as those over 60 years and who now constitute over 20 per cent of the population.)

Exactly the opposite is happening in the developing world. Here the number of babies born who survive into adulthood and themselves go on to reproduce is causing the logarithmic growth in population forecast by Malthus and his followers.

Family planning

There are so many factors that determine how many children a couple have that it is impossible to consider them all here. The drive to have sexual intercourse is equalled only by the urge to eat. Women are fertile for about 25 years of their life. During that time they can have 20 or more babies. In the population of French immigrants to Canada, in the 17th century, the average woman had 4.2 babies that survived to adulthood. This led to the 50,000 immigrants who remained there becoming the 6 million French Canadians resident in Quebec in the late 1900s when a survey was carried out.

Attempts to limit the number of children in a family have been made since records began. Infanticide and abortion were early crude attempts at limiting the size of a family. The earliest methods of contraception used both occlusion devices and chemical methods. Some of these, such as the use of an intra-vaginal sponge soaked in vinegar, are still in use. James Stuart Mill, in the 1821 edition of the *Encyclopaedia Britannica*, wrote, 'The grand practical problem, therefore, is to find the means of limiting the number of births.'

Effective contraception became available only through the efforts of a few valiant pioneers working to change people's attitudes in the early part of the 20th century. New York State made birth control a criminal offence in 1869 as a result of pressure from a group calling itself the Society for Prevention of Vice. The Comstock Law, enacted by Congress in 1872, made it a criminal offence to import, mail or transport in interstate commerce 'any article of medicine for the prevention of conception or for causing abortion'.

In 1906, the editor of a newspaper in Chicago was condemned to three years' hard labour for publishing a discussion on matrimonial relations in which contraception was mentioned. Ten years later, Margaret Sanger, probably the most famous campaigner for birth control of all time, was sentenced to serve 30 days in the workhouse for distributing a leaflet entitled 'Family Limitation'.

Things were not much better in other developed countries. Not that the methods of contraception available to them were very effective. Condoms were the most popular method of contraception up to the invention of the pill. They had originally been made from animal intestines, were expensive and used largely, if not entirely, by the rich as a prophylactic against syphilis and other venereal disease. Replaced by latex devices in the 1920s, they became generally more available. They were never very effective as a contraceptive and were confidently expected to be replaced by the pill. No one foresaw the AIDS epidemic that would increase the demand for condoms to prevent infection.

Early birth-control clinics advocated the use of occlusive devices such as malleable caps or discs that could be inserted into the vagina with or without a spermicide shortly before intercourse. These were favoured because it gave women control over their fertility rather than leaving it to their partner either to wear a condom or indulge in *coitus interruptus*. This practice, described in the Bible and generally referred to among its practitioners as 'being careful', was the only form of contraception apart from abstinence available to the poor. It is worthless.

Only with the development of the oral contraceptive did family planning gain respectability in the West and become a subject suitable for polite conversation. Voluntary sterilisation, which can be carried out on either partner, is a safe alternative. It is currently a surgical procedure and is comparatively cheap to perform, but it is impractical on the massive scale required to produce a significant effect on world population. As Professor A S Parkes, one of the pioneers of fertility studies, pointed out 40 years ago when the population was much smaller than today, it would take a thousand surgeons working eight hours a day, five days a week, some eight years to sterilise all the eligible men in India.

Why limit the size of families?

Even if contraception had been available to them, it is doubtful that our forefathers would have used it. The few children who survived to adulthood were an asset; they contributed to the prosperity and wellbeing of the family. The situation changed once more children survived than died in infancy and early childhood, as happened in most of the developed world during the 19th century. The Industrial Revolution made child labour unnecessary to create wealth and, with a more humanitarian approach to life, children became a (financial) burden rather than an asset.

Although advances in hygiene and medicine have greatly reduced infant and childhood mortality in the developing world, wealth creation has not occurred to the same extent. Most governments in the developing world have not seen it as necessary or desirable to advocate family planning, despite the increased infant-survival rates. Many sanctimonious observers in the West have castigated countries, such as China, that have realised the importance of limiting population growth, while condoning others whose citizens continue to multiply at a rate that inevitably reduces their standard of living and quality of life.

Abortion

Despite the effectiveness of barrier methods of contraception and of the contraceptive pill, women throughout the developed world continue to use abortion to terminate unwanted pregnancies. Although it was illegal in almost every country in the world in the past, abortion was widely practised as a means of limiting family size. It is still banned in countries that are in effect theocracies.

Currently, about 26 million women obtain legal abortions each year worldwide, while an additional 20 million abortions are estimated to occur in countries where it is restricted or prohibited by law. Close on 23 per cent of all pregnancies were aborted in England and Wales during 2007. The proportion in the USA is similar but varies enormously between states being as low as 2 and 4 per 1,000 women in Wyoming and Idaho and as high as 37 and 39 per 100 women in New York and California respectively. Abortions end around 30 per cent of all pregnancies in China. This has not changed much in the past half-century. It is not very dissimilar from those of other developing countries. It is estimated that 3 million abortions a year occur in Brazil, which confirmed its opposition to legalisation of abortion in 2008.

In Western society, abortion is more common among the underprivileged and less fortunate members of the community than among the affluent. Despite liberal sex education and free access to contraception, abortion remains an important factor in limiting population growth in the developed world.

A subtle modification of abortion as a means of family limitation is foetal female destruction by those who can afford it. This has led, in the middle classes in India, to a boy–girl ratio of 7:3. The practice has caused outrage but if it were to become commonplace it would eventually lead to the negative population growth the world so desperately needs.

The question arises as to what drives the high birth rate in the developing countries. Is it poor infant survival that leads to an increased number of pregnancies or is it that the increased birth rate results in poverty, which causes a high infant death rate? Probably both are true. The fact that women are prepared to risk their lives having abortions in unhygienic conditions favours the latter, whereas the low take-up rate of birth-control measures favours the former.

It is ironic that enormous amounts of money are spent on enabling the 10 per cent of women who are infertile to have babies in the developed world while in other parts of the world little is spent on helping them to control unwanted pregnancies. Are the two related? Probably not in the isolated cases we are used to, but it must have an effect in the developing world.

Does population growth matter?

The Industrial Revolution, without which modern life would be impossible, was confined, for much of its first 250 years, to the developed world. During the first half of this time, the population expanded rapidly but the rate of growth fell progressively during more recent times, until today it is below replacement levels in many European countries. This would undoubtedly have been a bad thing in former times, when virtually everyone was needed to produce the essentials for life, such as food, shelter, fuel and clothes. Once mechanisation made it possible for one man with a machine to produce as much food as ten men without one, an ever-increasing population was not required. Something similar happened to all human activities that involved manual labour. By the end of World War II, a favourite debating subject was what we would do with all our spare time.

In much of the undeveloped world, unemployment and poverty are rife and, as populations increase, it either remains static or gets worse. The situation is different in what have become known as the emergent nations, such as China and India. In these countries industrialisation is rapidly reaching that in the developed world, putting the finite resources on which they rely under severe strain.

Industrialisation brings with it an enormous rise in standard of living, though pockets of intense deprivation persist, even in the most developed countries. In the past, these improvement were largely at the expense of the undeveloped world, whose land, labour and mineral resources were exploited. In these countries, many lived in abject poverty. Today, because of modern telecommunications, their plight has become visible to millions in the developed world. As a result there is a humanitarian response to prevent some of the limiting factors to population growth – such as death from famines, wars and epidemics – from taking place. In the absence of voluntary or imposed birth control, and as a result of humanitarian intervention, their populations now increase at a rate comparable to that in the rapid-growth phase of the Western world.

No one can seriously believe that population growth can continue indefinitely. How large a population the world can sustain, and to what standard, is not and cannot be known with any certainty. One thing is certain, however: people use up precious finite resources and those in the developed world use more of them per capita than those in the developing world. Unless measures are taken to control the rate of population growth, there will be intolerable pressure on the supplies of food, water and energy. Even if scientific methodologies produce the means to feed and water this increasing population, it will be able to do so only by using up the diminishing supplies of finite resources, such as energy. Something must be done about the population explosion that is occurring in many parts of the world. We cannot or should not rely upon global famines, wars and epidemics to solve the problem.

The ethics and morality of population control

What is desirable or necessary for society as a whole is not necessarily consistent with what every individual wants. Even if every woman alive today had only two children that survived to adulthood, the world population will reach 9 billion within 30 years and remain at that level or higher. Few people believe that the world is large enough, or endowed with sufficient resources, to provide a population of this size with a standard of living much above subsistence level. The irony of the Pope, during a visit to Australia in 2008, urging governments to do something about CO_2 production without, at the same time, pointing out the dangers of the population explosion will not have escaped the attention of those concerned. Nor will those concerned about the future of our planet forgive the reluctance of governments to overcome the problem of political correctness and to face up to the need for population control in both the developed and developing worlds.

Epilogue

WHY DO WE BELIEVE IT?

STANLEY FELDMAN & VINCENT MARKS

DOGMA
It must be true: it was in the papers.

BEFORE THE AGE of Enlightenment, people believed what their Church told them. The few who questioned the word of the Establishment were labelled heretics. Today, we pride ourselves that it is evidence and rational thought that decide what we take to be truths. It is obvious we delude ourselves. While not everyone believes that all of the tall stories we are told are based on good evidence, so many people do buy into them that the consensus they create makes them credible. There is perhaps an excuse when the mistaken beliefs emanate from a democratically elected government – though even these may be suspect – but many of the canards that find considerable popular support come from dubious or self-interested but influential groups.

The government told us there were weapons of mass

destruction in Iraq. We preferred to believe the government rather than the expertise of the UN weapons inspectors. We had faith in our elected leaders, which proved to be misplaced but understandable. On the other hand, the so-called 'millennium bug' was obvious nonsense: there was no serious scientific evidence that the coming of the new century would cripple our computers, send aeroplanes crashing into one another or cause a worldwide breakdown in IT communications. Nevertheless, individuals, companies and government departments spent millions of pounds to make their computers 'millennium-proof'. How did such a silly scare take hold?

The salmonella-in-eggs story started with a basically correct statement by a government minister that was absurdly exaggerated until it exploded into a scare story that still reverberates in hospitals, schools and even among otherwise well-informed people who insist that all eggs should be hard-boiled to kill the harmful bacteria they might contain. Soon after the scare started, it became clear that, although it can occur, it is exceedingly rare for eggs to be infected by salmonella, and even rarer for anyone to suffer from it – but the damage was done.

Millions of pounds are spent, often by people who can ill afford it, on organic food in the belief that normal farming methods contaminate food with 'chemicals'. This is a nonsense. All food is composed of chemicals. The roots of plants can absorb only water-soluble ones. Soil itself is made up of many different types of chemical. The only source of chemicals available to plants, apart from those in the soil, is surface contamination by herbicides or insecticides. Most of these can easily be washed off. Some chemical pesticides are even used by 'organic' farmers, since without them the food would be contaminated by harmful fungi. It has repeatedly been demonstrated that there is no nutritional or toxicological advantage in organic food, so why do people fall for this scam? Indeed, with the need to increase food yields to meet the ever-expanding population, it is irresponsible to indulge in an agricultural method that produces a 40 per cent smaller yield than is possible using modern, scientifically tested, safe methods.

Some 50 years ago, it was suggested that a high intake of cholesterol causes heart disease. This was never established by any well-conducted clinical trial, so why are low-cholesterol diets still being advocated by so many authorities that should know better?

There is a very high-profile campaign aimed at reducing our salt intake. The evidence that this is beneficial is, at best, equivocal. There is, on the other hand, positive evidence that too little can, in certain circumstances, kill. Why have so many manufacturers taken salt out of foods, only to have them added by the consumer? The answer lies in perception generated in the consumer, the strength of which is increased by the very fact of responding to it. Thus, a vicious circle is set up in which evidence of benefit or harm plays a very little role.

The movement to ban plastic bags is based, among other things, on erroneous information about the harmful effect of plastic residue on sea birds and mammals and their indestructibility. But so great is the need for self-flagellation that many shoppers have embraced the idea with zeal.

Why do so many people oppose nuclear power stations in the UK when, 25 miles across the English Channel, France produces 80 per cent of its electricity this way? Is this all about to change with the rise in oil and gas prices and the fear of global warming? Let's hope so.

Why, when we are faced with feeding an enlarging population, do we prevent the growing and importation of any GM food, apart from tomato purée, in Europe? Over 50 per cent of the world has been growing and consuming GM foods for ten years without a single hazard either to health or the environment being recorded.

Why do so many intelligent people opt for alternative therapies to cure their illnesses? They imbibe expensive water blessed by homeopathy and herbal remedies that would never pass the tests of efficacy required of normal medicines, and they embark upon silly therapies from coffee enemas to electromagnetic diagnosis and therapy.

The greatest enigma of all is that of the CO_2-and-global-warming story. The unproven hypothesis that CO_2 is the major factor in the current episode of global warming has been taken as a given by authoritative bodies throughout the world, including the UN, national governments and many august scientific organisations. Is this a further example of the power of fashion in science or a genuine cause for concern?

The atmospheric-CO_2 scare story, based, among other things, on the thickness of tree rings – suggesting that the world is on the edge of Armageddon – has been so successfully sold that it won Al Gore a Nobel Prize. No serious scientific debate has been allowed in public because, we are told, all scientists agree. Many possibly do; even more apparently do not. How has this come about and who gave Biblical authority to its prophets?

Is the Age of Enlightenment coming to an end at the gates of a world of spin, propaganda and advertising? Are governments to blame? Are the media at fault? Are schools failing to teach the need to examine evidence in order to distinguish belief from fact? The promotion of 'intelligent design' as a science in the USA and in some UK schools suggests that they may be.

In their analysis of the great salmonella scare, Richard North and Christopher Booker clearly put the blame on the lack of leadership by the government department involved. However, without the part played by the media and the general gullibility of the public, someone, somewhere, surely would have questioned that the few cases of salmonella infection associated with mayonnaise might have been due to other components of the meal.

Governments

Governments are advised by committees chosen mainly from people who have a vested interest in, or preconceived prejudice about, the cause under scrutiny. All committees justify their existence by reporting that something must be done to solve the problem they were set up to investigate; few, if any, committees decide that things are OK or advocate leaving the status unchanged. As a result, the recommendations of 'experts' invariably reaffirm their own prejudices or, more commonly, those of their committees' chairs and secretaries. The committee's findings are then reported back to a government anxious to be seen to be doing something. This results in its issuing statements that subsequently fail to stand up to critical scrutiny. By then the damage has been done. The decision not to order the wholesale slaughter of badgers to reduce the incidence of tuberculosis in cows, despite good scientific evidence, is a recent example of the effect of biased interested parties on the decision-making process.

Pressure groups

Pressure groups are experienced at leading the media astray by presenting them with biased, unbalanced, incomplete and at times deliberately inaccurate information. They see this to be their job; they earnestly believe that they need to whip up public anxiety or anger to achieve their justifiable aims. Their organisation has no obligation to present a balanced picture.

Remember the Brent Spa oil rig that Greenpeace assured us was full of toxic waste and was going to pollute the ocean if it was scuttled in the depths of the Atlantic? The rig ended up in a Norwegian fiord, where, rather than pollute anything, it has become a breeding ground for fish and marine life. It did not stop Greenpeace presenting themselves as saviours of the seas. So it is that we get pictures of the Arctic ice breaking up without pointing out that it always does so in the summer months when the sun shines continuously day and night; of seabirds whose deaths are inaccurately attributed to plastic bags; of glaciers melting due to warming without pointing out they have been doing the self-same thing for hundreds of years.

Environmental campaigners do not believe that sending out rubber dinghies to throw paint bombs at whaling ships will stop them, or that killing whales for food will cause their extinction, but they are conscious of the David-and-Goliath picture it generates and how these affect public opinion.

Pictures of starving children in Africa soften the hardest hearts; sympathy and donations pour in and the pressure group moves on to the next calamity. The fact that Africa has been a net exporter of food is not mentioned, and nor is the fact that, in most cases, less than 30 per cent of any contribution to an NGO actually ends up helping to feed a starving child. This contrasts with 50–60 per cent of government assistance that goes directly to the children.

Antivivisection pressure groups produce lurid stories and pictures of animals being tortured. These bear no relation to experiments that are actually carried out nowadays under conditions that are closely monitored and controlled. Experimentalists are not monsters. Quite the contrary. They just happen to value human life more than that of animals. They are almost invariably more humane than their detractors, who are not beyond adopting terrorist tactics to achieve their misguided aims; and they are often among the first to volunteer to participate in experiments on themselves whenever this is possible

The media

Perhaps we should blame the media – the press, radio and television – for the silly things we do. It is they who take up the stories fed to them by pressure groups and, by making it news, cause it to have an influence on the reader, listener or viewer. In the process, any deceit or subterfuge becomes sanitised. What is a tendentious press release becomes, in many people's minds, an established fact.

Unfortunately, few reporters are trained in science or understand statistics. Reporters are seldom in a position to analyse critically a press release – the time constraints are just too strong. They are there to give readers what they want, which is a story. They frequently embellish the underlying story by presenting the most lurid or extreme examples that are more likely to grab the reader's attention. Their editors make no attempt to balance the picture with alternative views, as this would reduce the impact of their contribution. Not infrequently, the editorial policy on such matters as global warming, population control and nuclear energy has been made clear and reporters are under pressure to slant any piece to fit that policy.

The byline writer's job is to write a heading that demands reader attention. Too often it does not accurately reflect the actual content of the article but, nevertheless, it is the byline that is remembered. Some reporters, who say they would rather an *inaccurate* headline make a reader look at their contribution than an *accurate* one caused the story to be left unread, defend this policy, which, though unprofessional, is very common.

Whenever reporters are challenged, they defend their writings as 'giving the readers or viewers what they want'. Either they do not know they are being used as tools by various pressure groups or they engage in a prostitution of their media, conniving at misrepresentation for profit.

The public

At the end of the information chain come us, Joe Public. We are certainly not blameless. Unfortunately, most of us are uncritical and superstitious. Many of us do not walk under ladders if we can possibly avoid it; many of us are afraid of ghosts even if we do not believe in them; and we buy organic food in case the ordinary food that served our parents and us so well will harm our children. We are suspicious of technology we do not understand and opt for simplicity and nature. We believe that, if we nobly sacrifice something, it must do someone some good somewhere. This is the appeal of the environmentalists. At the end of the day, we must shoulder much of the blame for credence given to so many silly ideas. We become irrational, uncritical believers as soon as we are told our security or our health is threatened. Unfortunately, our critical shortcomings prove that it is possible to fool most of the people most of the time.

APPENDIX

Global Warming

Al Gore (2005), *An Inconvenient Truth: The Planetary Emergency of Global Warming and What We Can Do About It* (Rodale Press, USA) (a 2007 film, *An Inconvenient Truth*, followed). Gore was the unsuccessful presidential candidate whose book and film prophesied the doom and destruction of the planet due to manmade CO_2. He was awarded the Nobel Peace Prize in 2007.

IPCC (1990), 'Climate Change', scientific assessment reports of IPCC Working Group 1 (Cambridge University Press).

IPCC (1996), 'Climate Change: The Science of Climate Change' conference report, Working Group 1 (Cambridge University Press). There were a series of detailed studies by the IPCC Working Groups 1 and 2 between 1990 and 2000, which started with the proposition that manmade CO_2 was causing and would continue to cause global warming.

IPCC (2001) 'Climate Change', third assessment of climate change by the Intergovernmental Panel on Climate Change, The Scientific Basis (Cambridge University Press). There was a series of reports from this body on the socioeconomic impact, adaptation and the basis of the models used for predicting the future. Some of the predictions made in these reports have been used in this book.

A B Robinson, N Robinson and W Soon (2007), 'Environmental Effects of Increased Atmospheric CO_2', *Journal of American Physicians and Surgeons* 12(3), pp. 79–83.

Kyoto (1997), 'Kyoto Protocol to United Nations Framework Convention on Climate Change', online at:
http://edition. cnn.com/SPECIALS/1997/global.warming/stories/treaty.

APPENDIX

F Singer (2003), *The Revelle–Gore Story* (Hoover Press). Gore says it was Professor Revelle who introduced him to the dangers of CO_2 on the climate when he was at Harvard. Revelle was in fact a doubter; he wrote, 'The scientific base for a greenhouse warming is too uncertain to justify drastic action...' Gore supporters tried to suppress his views expressed in *Cosmos*, 2003.

R Lindzen (2003), *Global Warming*, the origin and nature of the alleged consensus, OPEC seminar on the environment (Cato Institute). Lindzen is an outspoken critic of the CO_2 global-warming theory. According to Lindzen, his colleague at Massachusetts Institute of Technology, Professor Newell, had his paper challenging the theory denied publication on the basis that the results were 'a threat to humanity'. Lindzen was to become involved in controversy when Gore said he had changed his views – he had not. (See Singer F.)

IPPC (2007), 4AR of Working Group 1, assessment report of the IPCC 2007 meeting held in Bali. This report contains the controversial Chapter 9, which presents manmade CO_2 as the cause of global warming as a fact. Several reviewers tried unsuccessfully to introduce a cautionary 'maybe' but were ignored. Much of the value of the report is compromised by the unconditional acceptance of the link as a proven fact.

N A Morner (2003), 'Estimating Future Sea Level Changes from Past Records', in 'Global and Planetary Change 40', summary for policymakers, Working Group 1.

A Rorsch (2008), 'Pseudoscientific elements in climate change research', online at http://www.ScienceandPublicPolicy.org.

D Holland (2007), 'Bias and Concealment in the IPPC process: The Hockeystick affair and its implications', *Energy and Environment* 18, p. 951.

Survey of Anesthesiology, 22, p. 419 (1978), 'Mathematical modelling is a pop art form. It is a pseudoscience converting present solid information into fallible predictions' Editor S Feldman.

B Lomborg (1998), *The Skeptical Environmentalist* (Cambridge University Press). This is an analysis of the statistical evidence for climate change.

B Lomborg (2007), *Cool It* (Marshall Cavendish). This is an analysis of the economic consequences of climate change.

C C Horner (2001), *Global Warming and Environmentalism* (Regnery Publishers, Inc.). This is a critical look at what has been claimed by the environmental lobby.

F Pearce (2006), *The Last Generation* (Eden Project Books). This looks at what might happen if the most alarmist predictions of the environmental lobby are right.

H Svensmark and N Calder (2007), *The Chilling Stars* (Icon Books). Svensmark has published extensively on the effect of cosmic rays on solar energy reaching Earth. These authors contributed to the film *The Great Climate Change Con*, made for Channel 4. This programme was referred to Ofcom who found they had been generally fair.

P Stott (2005), 'Global warming not a crisis', ABC News. Professor Sir Philip Stott has maintained that the South American rainforests are 87 per cent intact and the other 12 per cent is recovering.

L J D Haigh (2003), 'The effects of solar variability on the earth's climate', *Philosophical Transactions of the Royal Society*, volume A. 361.95.

A B Robinson, N E Robinson and W Soon (2007), 'Environmental Effects of Increased Atmospheric Carbon Dioxide' *Journal of American Physicians and Surgeons* 12(3), p. 79.

Nigel Lawson (2008), *An Appeal to Reason* (Duckworth).

Greenhouse Gases

Baron Joseph Fourier, *The Analytical Theory of Heat*. Fourier (1768–1830), a French mathematician, proposed the Earth's atmosphere protected the planet from the extreme heat of the sun.

Svante Arrhenius (1859–1927), a Swedish physicist, won the Nobel Prize in 1903 for his work on conduction of electricity in electrolytes.

Svante Arrhenius (1896), 'On the influence of Carbonic acid in Air upon the Temperature', *Journal of Science* 41, p. 237. This was the paper that started the greenhouse-gas story.

J Tyndall was an Irish physicist who described the Tyndall effect caused by refraction of light.

M Milankovitch (1941), *Canon of Insolation of the Ice-Age Problem* (English translation 1964). Milankovitch described the elliptical orbit of the Earth and the cyclical events that affect the temperature.

G S Callendar (1938), 'The Artificial production of CO_2 and its Influence on Climate', *Royal Meteorological Society Quarterly Journal* 64, p. 223.

Norman Lamont, Chancellor of the Exchequer in John Major's government, claimed in *The Great Climate Change Swindle* on Channel 4 in 2007 that the CO_2 story started with the aim of making nuclear energy more publicly acceptable.

APPENDIX

The History of the Atmosphere

S Chang (1994), 'The Planetary Setting of Prebiotic Evolution', part of the Nobel Symposium on *Early Life on Earth* (ed. S Bengst) (Columbia University Press).

H D Holland (1990), 'Origins of breathable air', *Nature* 347(17); also in *Nobel Symposium on Early Life on Earth*, 1994.

D Goldsmith (1997), 'Comet origin of oceans all wet?', *Science* 277, p. 318.

J F Kasting (1993), 'Earth's early atmosphere', *Science* 259, p. 920.

E G Nisbet (1991), *Living Earth* (Chapman & Hall).

R A Berner (2003), 'The long-term carbon cycle, fossil fuels and atmospheric composition', *Nature* 426, p. 323.

J F Nunn (1996), 'Evolution of the atmosphere', *Proceedings of the Geological Association* 109, p. 1.

G Walker (2003), 'Ice Magic', *New Scientist* 12, 30 April. *Snowball Earth*, Walker G, Crown, 2003, BBC 2 programme *Snowball Earth*, 2001.

N Calder (1997), *The Manic Sun* (Pilkington Press).

The Global Ice Caps

H J Zwally et al. (2005), 'Mass changes of the Greenland and Antarctic ice sheets and their contribution to the rise in sea level', *Journal of Glaciology* 51, p.509. They found net rise in sea level that could be attributed to melting ice caps to be in order of 0.5 mm/year.

Some of the information on the Arctic and Antarctic ice came from reports from NASA, which released the information from the Goddard Institute, 2006–07, headed by Dr Hansen, one of the IPCC founders, and from the European space satellites, European research satellite probes 1 and 2 (ERS 2007 University of Bremen). The level of the oceans has been plotted by ERS 1 and 2 by triangulation monitoring over 7–8 years.

Worldwide Fund for Nature, 'Polar Bears to be extinct by 2012'. In fact, numbers increasing in spite of about 500 being shot each year.

M Taylor and M Downsey (2008), in 'Demographic and Ecological Perspectives on the Status of Polar Bears' at:
http://scienceandpublicpolicy.org/.

A J Berner et al. (2005), *Arctic Climate Impact Assessment* (Cambridge University Press).

J Kohler 'Glaciology: lubricating lakes', *Nature* 445, p.904.

R E Bell (2008), 'Unquiet Ice', *Scientific American*, February.

332

The Economics of a Warmer Earth

R A Pielke (2005), 'Global Warming and Hurricanes', *Bulletin of the American Meteorological Society* 1571, p.75.

Leading Article (2007), *Daily Telegraph*, 3 December.

Lord Stern (2006), 'Stern Report'. This report presents models of the economic impact of global warming. The study, by a distinguished economist for the UK government presupposes a rise in global temperatures and in sea levels between two extremes based on the IPCC report of 2001. (This report was criticised by the economist Professor R Tol of Dublin in BBC Radio 4 broadcast 2008.) It is his predicted effect of a possible large rise in sea level at the top of the range of possible temperature effects, that have been reported in the media.

B Lomborg, *Cool It* (Marshall Cavendish). This book is largely devoted to the cost-effectiveness analysis of implementing the Kyoto Protocol.

Thomas Malthus (1798), *An Essay on the Principle of Population, as it Affects the Future Improvement of Society with Remarks on the Speculations of Mr Godwin, M Condorcet, and Other Writers* (Johnson).

A S Parkes, *New Scientist*, 27 July 1967, cited by P Ehrlich (1971), *The Population Bomb* (Pan Books).

Leslie J Reagan (1997), *When Abortion Was a Crime: Women, Medicine, and Law in the United States, 1867–1973* (University of California Press);

http://ark.cdlib.org/ark:/13030/ft967nb5z5/

http://www.abortion.com/Misc/AbortionStatistics.htm

BBC World Service, 10 July 2008

FURTHER READING

An Epidemic of Obesity

The Times, headline, 3 September 2008, p 4.

T A Welborn et al (2000), 'Body mass index and alternative indices of obesity in relation to height, triceps skinfold and subsequent mortality: the Busselton Health Study', *Int J Obes Relat Metab Disord* 24(1), pp 108–15.

R Andres (1985), 'Mortality and obesity: The rationale for age-specific height-weight tables', in R Andres, E L Bierman and W R Hazzard (eds), *Principles of Geriatric Medicine* (New York: McGraw Hill).

R P Troiano et al (1995), 'The relationship between bodyweight and mortality: a quantitative analysis of combined information from existing studies', *Int J Obs Relat Metab Disord* 20(1), pp. 63-75;

R A Drazo-Arvizu et al (1998), 'Mortality and optimum body mass index in a sample of US population', *American Journal of Epidemiology* 147(8), pp 739–49;

J T Gronninger (2006), 'A semi-parametric analysis of the relationship of body mass index to mortality', *American Journal of Public Health* 96(1), pp 173-8.

H Thomas (2006), 'Obesity prevention programs for children and youth: why are their results so modest?', *Health Education Research*, 21(6); 783-795

K M Flegal et al (2005), 'Excess deaths associated with underweight, overweight and obesity', *JAMA* 283(15) 18611867.

Y Takata et al (2007), 'Association between body mass index and mortality in an 80-year-old population', *Journal of the American Geriatrics Association* 55(6), pp 913-17.

P Diehr et al (2008), 'Weight, mortality, years of healthy life, and active life expectancy in older adults', *Journal of the American Geriatrics Society* 58(1), pp 76–83.

J Bailey et al (1986), 'Immunoreactive gastric inhibitory polypeptide and K cell hyperplasia in obese hyperglycaemic (ob/ob) mice fed high-fat and high-carbohydrate cafeteria diets', *Acta Endocrinol* 112, pp 224–9.

Electricity and Leukaemia

www.robedwards.com/2005/11/cancer_fears_ov_1.html

http://archive.recordonline.com/archive/2006/06/20/opinion-20views-06-20.html

Times Herald-Record, 20 June 2006.

www.epolitix.com/EN/Forums/CPIELFEMF

Leslie Hannah (1979), *Electricity before Nationalisation* (London: MacMillan).

Stephen Spender (1933), *Poems*.

National Grid EMF, online at www.emfs.info

www.iop.org/EJ/abstract/0952-4746/21/4/611

www.icnirp.de/documents/EPIreview1.pdf

www.time.com/time/magazine/article/0,9171,911755,00.html

http://en.wikipedia.org/wiki/New_York_City_Blackout_of_1977

http://query.nytimes.com/gst/fullpage.html?res=9B0DE2D91231F93AA35754C0A961948260

www.who.int/entity/pehemf/meetings/southkorea/en/Leeka_Kheifets_principle_.pdf

www.who.int/peh-emf/

www.icnirp.de/documents/EPIreview1.pdf

www.iarc.fr Volume 80 (2002), 'Non-Ionizing Radiation, Part 1: Static and Extremely Low-Frequency (ELF) Electric and Magnetic Fields'.

Health Protection Agency (2006), 'The distribution of childhood leukaemia and other childhood cancer in Great Britain 1969-1993', online at http://www.comare.org.uk/

The *Herald*'s figures for risk can be found in the abstract: www.bmj.com/cgi/doi/10.1136/bmj.330.7503.1290

www.hpa.org.uk/webw/HPAweb&Page&HPAwebAutoListNameDesc/Page/1158945066501?p=1158945066501

www.icnirp.de/documents/emfgdl.pdf

http://www.iegmp.org.uk/report/text.htm

'Watching the Directives: Scientific Advice on the EU Physical Agents (Electromagnetic Fields) Directive', online at: http://science_and_technology_committee/scitech290606.cfm

FURTHER READING

The group membership and links to its reports can be found on the website
www.rkpartnership.co.uk/sage
www.ehib.org/emf/RiskEvaluation/ExecSumm.pdf
www.hpa.org.uk/web/HPAweb&HPAwebStandard/HPAweb_C/12042766
82532

The Next Great Plague

www.hpa.org.uk
www.dh.gov.uk/en/Publichealth/Flu/PandemicFlu/index.htm
www.defra.gov.uk/animalh/diseases/notifiable/disease/ai/index.htm
www.who.int/csr/disease/avian_influenza/en/index.html
www.ukresilience.info/latest/human_pandemic.aspx
www.cdc.gov/flu
www.pandemicflu.gov

Unsustainable Arguments

'The Deluge of Disastermania', *Time*, 5 March 1979.

Christopher Lasch (1978), *The Culture of Narcissism: American Life in an Age of Dominishing Expectations* (Norton), p 3. The authors state that their 'process of determining behavior modes is "prediction" only in the most limited sense of the word.'

H Donella Meadows, et al (1972), 'Limits to Growth', Club of Rome.

Bjørn Lomborg and Oliver Rubin (2002), 'Limits to Growth', *Foreign Policy* No 133 (November–December), p 42.

Jared Diamond (2005), *Collapse: How Societies Choose to Fail or Survive* (Allen Lane).

Anton Chekhov (2006), *Uncle Vanya* (originally published 1897) (Dodo Press).

Geoffrey Borny (2006), *Interpreting Chekhov* (ANU E Press), p 180.

John Vidal, 'Engines of Change', *The Guardian*, 14 May 2008.

Ross Gelbspan (2007), 'Beyond the point of no return', *Gristmill*, 11 December, online at http://gristmill.grist.org

Andrew Simms, 'The Final Countdown', *The Guardian*, 1 August 2008.

Cahal Milmo, '"Too late to avoid global warming," say scientists', *The Independent*, 19 September 2007.

Geoffrey Lean, 'Phase out coal and burn trees instead, urges leading scientist', *The Independent on Sunday*, 14 September 2008.

Dr James Hansen (2008–9), 'Tipping Point: Perspective of a Climatologist', *State of the Wild*, p 9.

George Monbiot (2008), *Heat: How to Stop the Planet Burning* (South End Press).

Chris Rose, 'Change the world: A twelve-step programme', *The Independent*, 18 April 2005.

Jared Diamond (2004), *Collapse: How Societies Choose to Fail or Survive* (Allen Lane).

Paul Hyett, 'Does environmentalism turn humanism on its head?', speech at 'Audacity' conference, Building Centre, 10 July 2000.

Jeffrey Sachs, 'Lecture 2: Survival in the Anthropocene', Reith Lectures, Peking University, Beijing, 18 April 2007.

C Brahic, 'Unsustainable development "puts humanity at risk"', *New Scientist*, October 2007. Thomas Malthus (1798), *On The Principle of Population*, vol 2, p 261, quoted in Frank Furedi, 'Population Control', *Spiked*, 18 June 2007.

Jonathon Porritt, *Ecologist Online*, April 2007.

Brandon Keim, 'Population Bomb: Author Tackles Cultural Evolution', *Wired*, 12 March 2008.

Board of the Millennium Ecosystem Assessment, 'Living Beyond Our Means: Natural Assets And Human Well-Being', Millennium Ecosystem Assessment, p 4.

George Monbiot, 'Save Us From Ourselves', *New Statesman*, 30 June 2005.

Rajendra Pachauri, quoted in 'Shun meat, says UN climate expert', Al Jazeera English, 7 September 2008, online at:
http://english.aljazeera.net/news/europe/

Rajendra Pachauri, quoted in Alda Medeiros, 'Lifestyle changes can curb climate change: IPCC chief', *Human Village* 16 January 2008, online at www.humanvillage.com

Fred Pearce, 'The Population Bomb: Has It Been Defused?', *Yale Environment* 360, 11 August 2008, online at www.e360.yale.edu

Mark Lynas, 'A Green New Deal', *New Statesman*, 17 July 2008: 'with a "war economy" social mobilisation harnessed, this time not towards fighting fascism, but towards heading off ecological crisis.'

Achim Steiner, quoted in James Kanter, 'UN issues "final wake-up call" on population and environment', *International Herald Tribune*, 25 October 2007.

Boris Johnson, 'Global over-population is the real issue', *Daily Telegraph*, 25 October 2007.

John Gray, 'Straw Dogs: Thoughts on Humans and Other Animals ', *Granta*, p 33, 2002.

James Lovelock (2006), *The Revenge of Gaia: Why the Earth is Fighting Back – and How We Can Still Save Humanity* (Penguin).

Jonathan Porritt (2007), 'If I was in government... Jonathon Porritt makes population his number one issue', *Ecologist Online*.

Mary Colwell (2007), 'Too big for the planet?', *Tablet*, 21.

Andrew Wasley, 'A green army', *Ecologist*, 1 June 2008, online at: www.theecologist.org

Charles Clover, 'Jonathon Porritt: Britain should have "zero net immigration" policy', *Daily Telegraph*, 6 June 2007.

Tony Curzon-Price, 'Painful birth of a new epoch of simplicity', *Spectator*, 4 June 2008.

Tom Crompton, 'Weathercocks and Signposts: The Environment Movement at the Crossroads', April 2008, online at: http://assets.wwf.org.uk/downloads/weathercocks_report2.pdf

Elizabeth Farrelly (2008), *Blubberland: The Dangers of Happiness* (MIT Press).

Efficiency or Effectiveness?

Which?, June 2008, pp 16–17.

Population

G Stix, 'Traces of a distant past', *Scientific American* 299(1), July 2008, pp 38–45.

http://geography.about.com/od/obtainpopulationdata/a/worldpopulation.htm

Thomas Malthus (1798), *An Essay on the Principle of Population, as it Affects the Future Improvement of Society with Remarks on the Speculations of Mr Godwin, M Condorcet, and Other Writers* (Johnson).

A S Parkes, *New Scientist*, 27 July 1967, cited by P Ehrlich (1971), *The Population Bomb* (Pan Books).

Leslie J Reagan (1997), *When Abortion Was a Crime: Women, Medicine, and Law in the United States, 1867–1973* (University of California Press); http://ark.cdlib.org/ark:/13030/ft967nb5z5/

http://www.abortion.com/Misc/AbortionStatistics.htm

BBC World Service, 10 July 2008